For now the poet can not die.
Nor leave his music as of old,
But round him ere he scarce be cold
Begins the scandal and the cry.
—Alfred, Lord Tennyson

At ev'ry word a reputation dies.
—Alexander Pope

MATHEMATICAL
SCANDALS

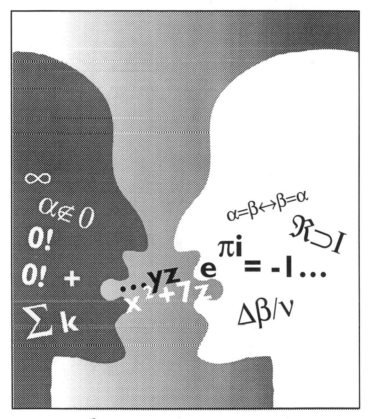

theoni pappas

Wide World Publishing/Tetra

Published by
WIDE WORLD PUBLISHING/TETRA
P.O. Box 476
San Carlos, CA 94070

4th Printing April 2002.

Printed in the United States of America

Library of Congress Cataloging-Publication Data

Pappas, Theoni
    Mathematical scandals / Theoni Pappas.
        p.   cm.
    Includes bibliographical references (p.   –   ) and index.
    ISBN: 1-884550-10-X   (pbk.)
    1. Mathematicians– –Biography.   I. Title
QA28.P36   1997
510' .92'2– –dc21
[B]   97-12297
                                                          CIP

Table of Contents

Introduction

Many people associate cold logic with mathematics. They consider the subject sterile and difficult to comprehend, and those that create it are often perceived as superior... nerdy... strange. Contrary to popular belief, mathematics is a passionate subject. Mathematicians are driven by creative passions that are difficult to describe, but are no less forceful than those that compel a musician to compose or an artist to paint. The mathematician, the composer, the artist succumb to the same foibles as any human — love, hate, addictions, revenge, jealousies, desires for fame and money. *Mathematical Scandals* is not meant to be a salacious book, but one that introduces the reader to the human sides and foibles of mathematics and mathematicians — illustrating that the mathematician is more than a theorem or a famous formula.

Each scandal is introduced by a vignette. Although these historical vignettes are fictional, they follow along factual lines of historical accounts. Bear in mind that in most cases the scandal is an almost insignificant yet fascinating part in the life of a mathematician aimed at revealing another facet of the person. Hopefully these stories will intrigue and tantalize you to search further and explore the mathematical works of some of these people. Who knows what surprises and twists are in store for you as you seek out additional information.

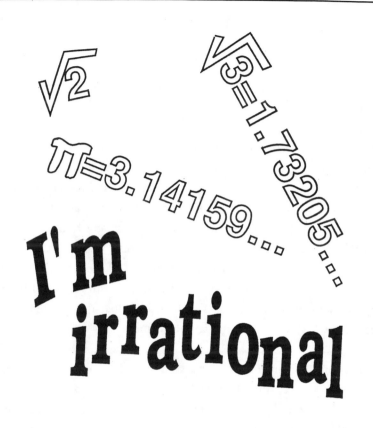

The irrational number cover-up

"Overboard with the traitor," the mob chanted.

"I am not a traitor to myself," Hippasus countered.

"You took the oath of the Pythagoreans. You have betrayed that vow, Hippasus," the leader of the group declared.

"I proved the unthinkable — the existence of irrational numbers. You want me to keep this a secret? You are asking me to suppress knowledge and truth, " Hippasus said boldly.

"You know we said that they are not numbers," the leader replied.

"But $\sqrt{2}$ is a number! Is not the function of a number to measure? $\sqrt{2}$ measures a specific length. No other number can give the exact magnitude of the diagonal of a 1 by 1 square," Hippasus insisted.

The crowd of Pythagoreans on board the ship were becoming increasingly impatient. The truth was upsetting them. Suddenly, their cries were followed by action. Everything happened very quickly. No one aboard could stop the momentum of the mob. "Overboard with him" they shouted as they sought to conceal the unconcealable. $\sqrt{2}=1.41421\ldots$ They grabbed Hippasus and threw him overboard to his death.

And so it was on a sea voyage that Hippasus was confronted for revealing and taking personal credit for the $\sqrt{2}$ secret. The rage of those on the boat could not be contained. They punished the "traitor".

Cover-ups are sometimes considered a 20th century phenomena — with such infamous ones as *Watergate* and *Iran- Contra* topping the list. History has many other examples, but who would think a mathematical cover-up would exist? Why would one want to conceal the discovery of new numbers? Until Hippasus' proof, the Pythagoreans believed that only whole numbers and their ratios could describe anything geometric. Although no one had found a fraction[1] that would give the exact number description for the diagonal of a 1x1 square, it was believed that its ratio of whole numbers had not yet been found. The Pythagoreans could not accept the need for the existence of other numbers. But when

2

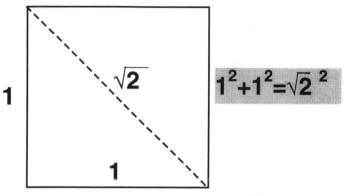

The diagonal of a unit square is the $\sqrt{2}$.

Hippasus showed that no other existing number could express this square's diagonal, the Pythagoreans were disturbed, to put it mildly. In fact, they contended that such numbers as $\sqrt{2}$ were not numbers.

A shroud of secrecy and mysticism engulfed the society of Pythagoreans. Unlike an ordinary school, the Pythagoreans had many restrictions that had to be followed. They were sworn to secrecy. Ideas discovered were not to be credited to the individual but to the Pythagoreans in general. Written records of their beliefs and their discoveries were prohibited.

Mathematics played a very unusual role in their lives. It was a philosophy of life that impacted their entire belief system. Their edict was *All is number.* To them the essence of the universe was numbers and in particular whole numbers and their ratios (fractions). The Pythagoreans used whole numbers and fractions to describe everything from people to music. 1 was the sacred creator of all numbers, since any whole number was the sum of a finite num-

PITAGORAS

Etching illustrating Pythagoras working with whole number ratios and musical notes.

ber of ones. 2, the first even number, was considered a female number and was associated with diversity of opinion. The first male number was 3 which was linked to harmony since it was a combination of 1 and 2. 4 stood for justice. 5 was linked to marriage, since it could be the union of 2 and 3. On and on they personified numbers with such names as amicable, perfect, abundant, narcissistic.

They considered the whole numbers the rulers. The Pythagoreans believed that every other type of number that existed could be expressed as a ratio of whole numbers. Life with these numbers was well ordered, and clearly described their world. *Enter the proof of the famous Pythagorean theorem* — the theorem that would immortalize the name of the Pythagoreans. Enter the decline of the superiority of the whole numbers. With the Pythagorean theorem the whole numbers' rule came crashing down.

Our imaginations are left to conjure up the feeling of the Pythagoreans on board when they learned that Hippasus had betrayed their oath of secrecy, and he declared he could proved that not all numbers could be expressed as whole number ratios. Imagine his feeling upon discovering and proving the irrationality [2] of $\sqrt{2}$. Imagine being able to see the specific length of $\sqrt{2}$ (the diagonal of a square)   yet not being capable of expressing it precisely by the numbers which had molded and dominated the Pythagorean belief system. Imagine the look on Pythagoras' face when the idea was presented to him by one of his followers. The gut reaction must have been — "It can't be" — "We must not let this out." Could their oath of secrecy conceal this?   How long would the cover-up last? How could a discovery so significant not be disclosed? Perhaps a non-Pythagorean would stumble upon it, especially since knowledge of the Pythagorean theorem had been known for centuries in many parts of the world.

There are numerous accounts of the above story.   All agree that Hippasus of Metapontum did prove the existence of irrational numbers in the 5th century B.C. and that he was expelled from the Pythagorean society. Some accounts record that *he was put to death at sea.* Others say *he was expelled from the society, and the Pythagoreans made a mock grave with a tombstone marking his "death".*

The secrecy of the Pythagorean society makes it impossible to know the exact means or reason of Hippasus' expulsion. Here are some of the various possibilities.

Was he punished and expelled because—

   • *he revealed his discovery of the irrational number $\sqrt{2}$, thereby breaking his vows of secrecy and individuality?*

   • *he was instrumental in heading a movement contrary to the rules of the conservative and secretive Pythagorean society?*

   • *he disclosed the findings of certain geometric figures, namely the pentagon and/or the dodecahedron?*

- *of a combination of infractions against this secret society with the last straw being the unveiling √2̄?*

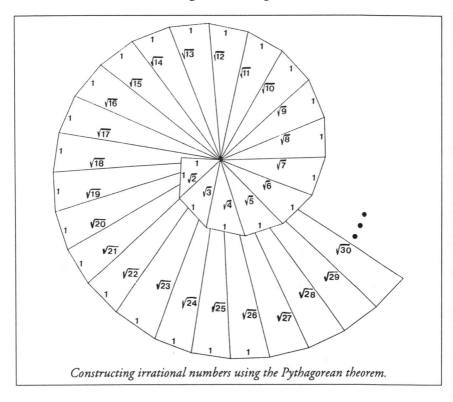

Constructing irrational numbers using the Pythagorean theorem.

[1] Fractions—ratios expressed by whole numbers— were called commensurable.

[2] Irrational numbers were known as incommensurables. In Greek they were called αλογος meaning inexpressible or αρρητος, which means not having ratio. Dealing with these incommensurable numbers was quite a problem, since they had not been clearly defined. The Greeks worked with irrational numbers by considering their size (magnitudes) in geometric terms rather than numerical terms. They could determine the exact magnitudes of many using the Pythagorean theorem, which itself was also stated in a geometric manner by constructing squares on the legs and hypotenuse. The Babylonians, on the other hand, had devised their decimal (sexagesimal) approximations for irrational numbers, but had no idea that exact decimal approximation for them did not exist.

Ada Byron Lovelace's Addiction

"No, John!" Ada raised her voice. "You mustn't tell William about us. I don't want to hurt him."

"What about me? What will I do about all the gambling debts we have?"

"You have my life insurance to cover them. My husband has paid my debts many times. My mother has paid the pawnbroker for the Lovelace diamonds I secretly pawned. Even though William now knows you are not a bachelor and have a family, he mustn't know about our relationship. Please," she pleaded. "What do you want from me that you don't already have?"

"I want the Lovelace diamonds. I can pawn them again," he answered.

"Why did I get involved with such a man? " Ada wondered to herself. "I thought of my marriage as a lifeless life. How could I have been so foolish? This is not the life I envisioned.

And why haven't I been able to conquer my gambling addiction? We thought we had a system. The gambling ring used my status to elevate themselves. And in the end we all came down. I should have listened to Greig's and Babbage's warnings. But I've always loved horses. To watch them race is a thrill. Add to this a bet and the excitement is intensified."

She went into the other room for a moment, leaving Crosse alone. Returning, she thrust out her right hand, saying "Here, take the diamonds. Pawn them again. But keep your silence."

* * *

Ada's thoughts raced over the years, recalling people and places. The cancer was consuming her body. Her time was short. Could she get her affairs in order? She had told her husband that she wished to be buried next to her father, Lord Byron. She wrote to Babbage asking him to be her executor, giving specific instructions on how to dispose of her estate, belongings, papers, and letters. After Ada's death, when her mother learned of Ada's letter to Babbage, she refused to acknowledge it. But Babbage would not let Ada's wishes be ignored and replied, "With respect to the collections of letters and papers given by Lady L. to Mr. B. during her life as well as an extensive correspondence carried on with her for years, many parts of which are highly creditable to her intellect, Mr. B. feels entirely at liberty to deal with them in any manner he may choose. The conduct of Lady L's relatives to Mr. B. has released him from the feeling of the delicacy towards them;

and Lady L's testamentary letter gives him full authority." [1]

*   *   *

"Mother, I need something for the pain. Why have you stopped the morphine and opium?"

"Dear, the pain will cleanse your soul," Lady Byron said, a strained smile on her face.

A rendition of a sketch done by her mother while Ada was on her death bed.

Today Ada Byron Lovelace is remembered for two things — she was the daughter of Lord Byron and she is considered the first computer programmer. What makes the latter amazing is that the computer she programmed had not been built. She had no way to test her programs other than studying and understanding how

9

Lord Byron(1788-1824)

Charles Babbage's difference and analytical machines would function. How she devised her programs and the events of her life are fascinating. She lived during a period when women were not encouraged to seek an education, especially one dealing with the sciences. Such fields as mathematics were considered harmful to a woman's health because they would tax her fragile brain. Fortunately, such beliefs did not discourage Ada's mother, Lady Byron, from studying mathematics, a subject she enjoyed. As a result she encouraged Ada's [2] mathematical interests.

Many doors were open to Ada because of her family's prominent position. As a result, Ada was able to converse, interact and study with a variety of talented people. Among these were Mary Fairfax Somerville, Augustus De Morgan, Charles Babbage, and Charles Dickens.

In 1835, Ada married William King[3] (11 years her senior), and in the next four years she gave birth to two sons and a daughter. But Ada was not satisfied with the role of nurturing mother. She wanted to pursue her ideas and studies, let her imagination soar, and explore ideas. Fortunately, the rearing of her children was gladly supervised by her mother and husband.

Because of her aptitude in mathematics and her fluency in French, the British journal, *Taylor's Scientific Memoirs,* asked her to translate Luigi Federico Menabrea's description of how Babbage's machines functioned and produced tables. She would later refer to her translation of the article for *Taylor's Scientific Memoirs,* as her "first born". The article had appeared in French in October of 1842 in *Bibliotheque Universelle de Genève.* Realizing Menabrea's article only dealt with the mathematical principles behind Babbage's machines, Ada decided to enhance the translation by adding her own information in the form of notes. As she labored over Babbage's plans and drawings of the Difference and Analytical engines, both analyzing and explaining the mathematics, she expounded on the machines' mechanics and uses. Babbage's machines were to be operated using punched cards analogous to the Jacquard-loom, but far more advanced. In her notes she pointed out many of the special features of the Analytical engine such as the *mill* where calculations would take place, the *storehouse* where results were stored, the *backing of cards* which allowed it to reuse any card or set of cards any number of times in solving a problem. *Remember, Ada was describing a machine that did not yet exist.* She explained how a Babbage machine could do a job with *three* cards that the Jacquard-loom would take 330 cards. She explained how the machines could tackle problems that were unsolved, e.g. astronomical tables, generate random numbers, compute complex sequences of numbers. In fact, she wrote a program for computing Bernoulli numbers, explaining how and where to set calculations and read results. This was an impressive task, especially when one realizes she had no machine to work out the bugs of her programs. Her notes ended up being three times the length of the translation, with insights and imagination that even Babbage said he had not considered. Ada was in her element when she explored ideas.

Perhaps it was the field of probability that got Ada into difficulties, or perhaps it was her fondness for horses. As a young girl Ada enjoyed riding and loved horses. Apparently the thrill of horseracing contributed to Ada's involvement in gambling. Having been introduced to mathematician John Crosse by his father, scientist Andrew Crosse, she was charmed by him. She wrote "...Young Crosse is an excellent mathematician...and he works my brain famously for he opposes everything I advance, intentionally; but with perfectly good humor...He will be an addition to my catalogue of useful and intellectual friends."[4] In the meantime, their relationship grew less and less platonic, and they soon became lovers. In addition to shared interest in mathematics, they discovered they both had an insatiable love for gambling, in particular horseracing. What might have started as a casual pastime became a compulsion for Ada. The more she got into debt, the more she tried to recoup her losses. By 1848 Ada was in such debt, that she asked her close friend Wonronzow Greig (Mary Sommerville's son) to help her arrange for a loan without her husband's knowledge. However, this money did not go far, and once more she found herself deeply in debt. She then asked her husband for an increase in her yearly allowance [5] claiming she had incurred her debts by excessive spending on books and music. Naturally she did not tell him about her gambling or the money she had spent on furnishing Crosse's home. After taking care of her debts, her husband decided to take her on a vacation. Unaware of the seriousness of his wife's addiction to horseracing, he made the mistake of allowing her to go to the Donaster races without him. Here, she became hooked again, but this time she had a letter indicating that Lord Lovelace would cover her bets. Her debts skyrocketed.[6]

Not in good health for some time, her condition progressively deteriorated even as her gambling became frantic and drastic. She became involved with a ring of gamblers with whom she had to arrange a life insurance policy specifically to cover her debts. She even

Ada at seventeen.

gave Crosse her husband's promissory debt letter, thereby becoming ever more desperate and deceitful. She had secretly pawned the Lovelace diamonds for £800 and replaced them with fakes. She eventually confessed this to her mother, who retrieved them for her. In 1852 Lovelace told Ada he had learned their friend Crosse was secretly married and had a family. Ada was concerned that perhaps Crosse would reveal the truth about them to Lovelace, so she bought Crosse's silence with the Lovelace diamonds. During 1852, the last year of her life, her cervical cancer progressed slowly and painfully. Her mother took charge of her care and her household, and Ada lost control of her life. During her last days, she confessed her affair with Crosse to her mother, but did not tell her she had secretly given him Lord Byron's gold ring, locket with Byron's hair, and miniature portrait of "Maid of Athens". In the past, Ada had been able to control her pain with morphine and opium, but now her mother withheld these medications so that Ada's pain would help redeem her soul. Her last days were filled with agonizing pain. In 1852, at the age of 36, Ada Byron Lovelace died. After her death Crosse claimed her life insurance, and sold the intimate letters she had written him to Ada's family. Lady Byron distanced both herself and Ada's children from their father. She held Lord Lovelace responsible for not stopping Ada's gambling and causing the rift between her and Ada. In addition, she was angry because Lovelace had respected his wife's wishes and buried her next to her father, Lord Byron.

The role Ada played in mathematics and more specifically computer programming did not come to full light until the mid-1900s when her work was found by a researcher. Today, Ada Byron Lovelace is recognized for her contributions to computer programming. In honor of her programming work the American National Standards Institute approved ADA as a national all-purpose standard and assigned it document number MIL-STD-1815, the year Ada was born.

The Difference Engine, one of Babbage's famous calculating machines and Charles Babbages (1792-1871).

[1]*Ada: A Life and a Legacy.* Cambridge, MA. MIT Press, 1985. p. 258.

[2] Lady Byron left Lord Byron three months after Ada's birth. After separation papers were drawn, Lord Byron left England.

[3] Later her husband became Lord Lovelace.

[4]*Ada Byron Lovelace* by Mary Dodson Wade. Dillion Press, NY. 1994 page 96.

[5]As her husband, Lord Lovelace retained control of her money. Her annual allowance was £300.

[6]At the 1851 Derby Day she lost £3,200.

Exposing L'Hospital's claim to fame

"No, Marquis, that is not what Leibniz intended," Johann Bernoulli explained. "Let me demonstrate this theorem." He then proceeded to explain a fine point from one of the recently published papers [1] of Leibniz.

The Marquis de L'Hospital had become intrigued with the new field of mathematics Leibniz was developing, but having only studied mathematics as a pastime he lacked the tools and skills to fully comprehend the new ideas of calculus. It was common knowledge that both of the Bernoulli brothers (Johann and Jakob) had been of invaluable assistance to Leibniz in his development of calculus. What better place to seek help with these new ideas than from one of these experts. The Marquis thus decided to ask Johann to tutor him. The very talented Johann viewed tutoring the Marquis as a source of income and an opportunity to make additional connections among the nobleman's circle of friends.

Johann Bernoulli 1667-1748

After working with the Marquis for a number of months, both at the Marquis' country chateau and his Paris home, it was time for Bernoulli to return to his home in Basel, Switzerland. The two men agreed to correspond, and exchange ideas and questions.

Working on his own and with the insightful letters of Johann, the Marquis came to realize his discoveries and ideas in this area of mathematics were insignificant when compared to those of Leibniz and Johann. The Marquis knew he was a very proficient but amateur mathematician. He lacked that certain creative and intuitive ability he had witnessed time and again in Johann. Yet he desired to make a name for himself in this subject he so loved. He wanted to be famous for something other than his noble title. Thus driven, he wrote the following on March 17, 1694.

> My dear Johann—
>
> It seems we both have need of one another. I need your intellectual help and you can use my financial help. Therefore, I propose the following agreement.
>
> I will give you a pension of three hundred livres this year. In addition, I will send you two hundred livres for the journals you have sent me. Since I realize it is a very moderate sum, I promise to increase the pension as soon as I get my affairs in order. I do not expect you to devote all your time to my needs, but ask you to set aside a few hours every so often to work certain questions and problems. In addition, I ask that you let me know of the new discoveries you might make, and that these not be communicated to anyone else. I also require that you not send copies to anyone of the notes you have sent me because I do not want them made public.
>
> I await your answer to this arrangement.
>
> Yours truly,
le M. de L'Hospital

Johann was a bit surprised when he received the Marquis' letter, but he had to be practical. He had recently married, and did not as yet have permanent employment [2]. Doing this for a short time would definitely be of help to him.

Johann guessed the Marquis would probably try to impress various people with his work by passing it off as his own. But Johann did not expect the Marquis would use his work in books under his own name. In 1696, L'Hospital's **Analyse des infiniment petits** was published in Paris. Many of the ideas of Bernoulli and Leibniz appeared in L'Hospital's book. The Marquis cleverly included the following disclaimer— "I have made free use of their discoveries, so that I frankly return to them whatever they please to claim as their own." His book became quite well known and was especially known for having the rule for the expressions of 0/0, which came to be known as **L'Hospital's Rule** [3] — the rule that put L'Hospital's name in mathematical history.

In the meantime, because of their agreement, Bernoulli felt honor bound that he could not reveal publicly what parts of the book were his until after L'Hospital's death. Even then it would be his word against that of L'Hospital's. He would not really receive credit for his work until 1955! To this day the rule is still called **L'Hospital's Rule**.

The truth about the relationship between L'Hospital [4] and Bernoulli had been unknown for hundreds of years. After L'Hospital died, Bernoulli felt he could make his discoveries public, which he did in 1704 in **Acta Eruditorum**. Because it appeared after

Marquis de G.F. A. L'Hospital, 1661-1704)

L'Hospital's death, however, it still left the issue cloudy. The first breakthrough in the true facts occurred in 1922 when Johann Bernoulli's *Lectiones de caculo differentialium* [5] (lectures he had given during 1691-92) were published which predated L'Hospital's book. Comparisons of both works imply that L'Hospital had lifted freely from Bernoulli. The final piece of evidence came to light in 1955 with the publication of *Der Briefwechsel von Johann Bernoulli* (edited by O. Spiess) which contains the correspondence (presented here in a colloquial translation) from L'Hospital to Bernoulli describing their agreement, their arrangement for payment and L'Hospital's exclusive rights to some of Bernoulli's discoveries. In addition, a letter dated July 22, 1694 from Bernoulli to L'Hospital contained the rule for 0/0, which predated L'Hospital's book!

> *L'Hospital's Rule*
> *Reveals the indeterminant form of 0/0.*
> *If the limit of the ratio of two functions,*
> *f(x) /g(x) is 0/0 as x—>a,*
> *then use the smallest derivative, d,*
> *of the functions*
> *for which $f^d(x)/g^d(x)$ is not 0/0.*

---

[1]The papers dealing with the calculus of Gottfried Wilhelm Leibniz were published in 1684 and 1686. It is well known that Leibniz had begun to collaborate with both Jakob and Johann Bernoulli in 1685 after the publication of Leibniz's first paper on calculus.

[2]In 1695 Johann Bernoulli became professor of mathematics at the University of Groningen in the Netherlands. His brother, Jakob, in the meantime occupied the chair of mathematics at the University of Basel. When Jakob died in 1705, Johann took over this position.

[3]L'Hospital's rule deals with the ratio of two functions each of which limit is 0 when x—>a, a is a particular value. The rule deals with finding the ratio using the derivatives of the functions.

[4]Guillaume François Antoine de l'Hospital (1661-1704) was a French nobleman, who enjoyed studying mathematics. Johann Bernoulli (1667-1748) was a member of a prominent merchant family in Basel, Switzerland. He, along with his brother Jakob (1654-1705), made many outstanding contributions to mathematics, especially to the field of calculus. In addition, Johann was a gifted teacher who was instrumental in spreading the power and appreciation of calculus on the European continent.

[5]His integral calculus book *Opera omnia* (Lausannem, Geneva, 1742) also added credence to his claim.

21

WHOSE solids ARE THEY ANYWAY?

"These five beautiful solids are a mathematical way of describing the elements — fire, water, air and earth — plus the universe," Plato pointed out to his friend. "In fact, I have written my thoughts on this topic in my dialogue Timaeus[1]."

"Do explain," his friend urged.

"These solids are as exquisite as the earthly elements they describe. Each is composed of a finite number of faces that are the same size and shape on all sides, and that mysteriously fit together. I have assigned each of the four elements a solid. Fire is the **tetrahedron**, earth is the **cube**, water is the **icosahedron**, and air the **octahedron**." Plato drew the solids as he spoke.

"But what of the **dodecahedron**?" the friend asked.

tetrahedron

hexahedron
(cube)

octahedron

dodecahedron

icosahedron

The five Platonic solids.

"That, my dear Dioclides, represents the universe. Its twelve faces represent the twelve zodiacs."

"Of course, I should have known. You are so clever to have thought of this link, Plato," the friend complimented.

*   *   *

And so, even though Plato did not discover any of the five Platonic solids, he described them in vivid detail in his Timaeus. It was in this dialogue he assigned their shapes to natural phenomena. Because of the fame his dialogues assumed over the centuries, the solids were incorrectly attributed to Plato and the myth perpetuated. Unlike others who deliberately tried to get credit for work that was not theirs, Plato would go out of his way to acknowledge another's works. But whose solids are they?

T he tetrahedron, the cube, and the dodecahedron were discovered by the Pythagoreans. Plato learned about various Pythagorean ideas on his travels to Greek cities in Africa and the Italian peninsula, especially in Sicily in 388 B.C.. The other two regular solids — the octahedron and the icosahedron are credited to the

Plato (c. 428-348 B.C.)

mathematician Theaetetus[2]. In addition, Theaetetus probably proved the theorem which appears in Euclid's *Elements* [3] — showing there exist five and only five regular convex solids. Plato was close to Theaetetus, and wrote a dialogue called *Theaetetus* as a tribute to his friend who had been wounded in battle and subsequently died. Plato had no intention of having these solids named after himself, but rather chose to sing their praises.

---

[1] Whether the Pythagorean Timaeus of Lorci was a real person is uncertain, but Plato used him to convey the ideas of the Pythagoreans in his dialogue *Timaeus*.

[2] Theaetetus (414 B.C.–369 B.C) came from a wealthy family in Attica, and made many mathematical contributions to the *Elements* including extensive work on the five solids and the theory of irrational numbers.

[3] For the proof see scholium No. 1 to Book XIII of Euclid's *Elements* or the Historical Note given on page 438 of volume 3 of *Euclid's Elements* translated by Sir Thomas L. Heath, Dover Publications, NY, 1956.

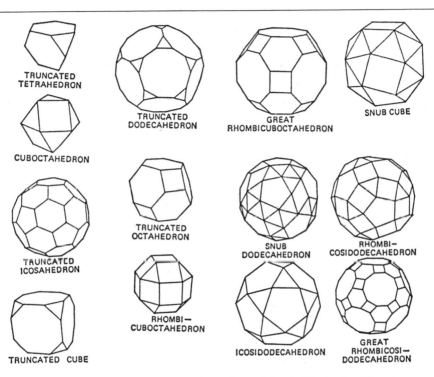

TRUNCATED TETRAHEDRON

CUBOCTAHEDRON

TRUNCATED DODECAHEDRON

GREAT RHOMBICUBOCTAHEDRON

SNUB CUBE

TRUNCATED ICOSAHEDRON

TRUNCATED OCTAHEDRON

SNUB DODECAHEDRON

RHOMBI– COSIDODECAHEDRON

TRUNCATED CUBE

RHOMBI– CUBOCTAHEDRON

ICOSIDODECAHEDRON

GREAT RHOMBICOSI– DODECAHEDRON

These non-Platonic are called Archimedean solids. Unlike the five Platonic solids, the faces on any one of these solids are not congruent to one another.

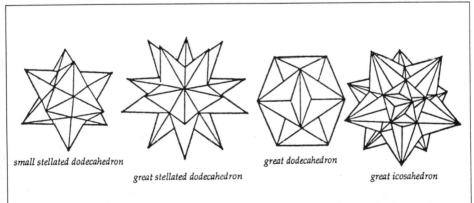

small stellated dodecahedron

great stellated dodecahedron

great dodecahedron

great icosahedron

Two of these non-Platonic solids were discovered by Johannes Kepler and two by Louis Poinsot. These solids are not convex.

25

The paranoia of Kurt Gödel

"I assure you, Oskar, it won't be a problem," Einstein said, trying to seem confident.

But Morgenstern[1] was not so sure — having just hung up from speaking with Gödel on the telephone. "Albert, he is insisting that he found a loophole in the constitution that would make a dictatorship possible in the United States. I insisted that was totally unlikely, what with all the checks and balances written into the document. But as you know his logic is impeccable. If he says he has found a loophole, he has found one. Above and beyond that, I told him that was not his problem at this point. His problem was to pass the interview for his citizenship papers. I asked him to not focus on this loophole — to not even to think about it, but just answer the simple questions of his interviewer."

"But you are right Oskar, nothing is just simple and straightforward for Kurt. He considers all facets and ramifications of a question. We must try to keep him from going off on this tangent," Einstein replied. "We must keep him distracted until the interview."

"For one so logical, he can be very impractical at times," Morgenstern added. "Common sense seems to sometimes get clouded in his brilliance."

* * *

At his citizenship hearing, the interviewing judge was star struck and invited Gödel's celebrity witnesses to be present. Einstein and Morgenstern were pleased, feeling they were now in a better position to ward off any problems that might arise during the interview.

"Until now you have been a German citizen," the judge commented to Gödel.

"No, I am Austrian," Gödel corrected the judge.

"Yes, of course," the judge replied. "Nonetheless, Austria has come under the dictatorship of Hitler. Fortunately, this sort of thing is not possible in the United States." The worst fears of Einstein and Morgenstern surfaced with that comment. The judge had opened Pandora's box.

"I can prove to the contrary," Gödel asserted, hitting his fist on the table. Gödel began his diatribe, which all three observers were hard pressed to terminate. After much effort, Gödel was calmed, and the "situation" was resolved. His citizenship was granted.

* * *

"Doctors... I cannot trust them, Adele," Gödel blurted out to his wife, while at the dinner table.

"Dear, you're not feeling well. You hardly ate," his wife replied.

"I am afraid something has been done to the food," Gödel countered.

"Dear, you trust me. Don't be frightened. I've prepared everything here." His wife had to constantly reassure him in this way whenever they sat down to eat.

His paranoia worsened when Adele had to have major surgery and then enter a nursing home to convalesce. She was not there to nurture and coax him to eat, to assure him the food was fine.

*   *   *

"I can't eat this," Gödel thought. "I'm sure it is tainted. Perhaps the butcher was paid to add something.

"The milk, the vegetables. Nothing, nothing, I mustn't eat anything. I know they are trying to poison me."

He was now frantic. His paranoia was poisoning his mind, his every action, his very thoughts.

During Adele's absence, Gödel did not eat. He became weaker with each passing day. Finally he had to be hospitalized on December 19, 1977. During the weeks in the hospital he refused treatment and food. He died there on January 14, 1978.

And so the door closes on the life of one of the most brilliant logicians of this century. He in essence had starved himself to death. What led up to his emotional troubles? Why was it so difficult for him to cope with everyday social situations?

T hese are but a few of the many bizarre stories that surround Kurt Gödel. During his lifetime Gödel was plagued by paranoia. He was a genius on the verge of insanity. One can only speculate about the cause of his paranoia. Perhaps the events of his life shed some light.

Born in 1906 into a German-speaking Lutheran family (part of the minority population in Brünn, Morovia), Gödel was the younger of two sons of Rudolf, a textile factory worker, and Marianne. Gödel's brother became a radiologist, and Gödel studied mathematics at the University of Vienna, earning his doctorate in 1930. At the Königsberg conference in 1930, he first presented his very novel ideas in a talk. The power and consequences of these ideas seemed to be beyond most of the attendees. If they had grasped the ramifications of what Gödel had presented, Königsberg's fame would have once more been revisited by mathematics.[2] In 1931 his paper, *On Formally Undecidable Proposition of **Principia Mathematica** and Related Systems* was published in a German mathematics journal. In this article he formulated and proved his now famous and far reaching *Incompleteness Theorem*[3] which placed insurmountable obstacles in the works of such renowned mathema-

Kurt Gödel (1906-1978)

ticians as Bertrand Russell, Alfred North Whitehead and David Hilbert. In fact, "...Gödel's Theorem shows that human thought is more complex and less mechanical than anyone had ever believed, but after the initial flurry of the excitement in the 1930s, the result ossified into a piece of technical mathematics... and became the private property of the mathematical logic establishment, and many of these academics were contemptuous of any suggestions that the theorem could have something to do with the real world."[4] Actually, the implications and beauty of Gödel's theorem are vast, applying not only to mathematical systems, but touching such areas as computer science, economics, politics, and especially nature. His theorem essentially says that not all true statements that evolve from an axiomatic system can be proven — we can never know everything nor prove everything we discover. In a broader sense, this idea extends and implies that our thoughts, ideas, the universe are infinite. Gödel's *incompleteness theorem* may seem obvious and true, but its beauty is that it is not an axiom (something we assume true) but rather a theorem (something *proven* true.) The task and methods of his proof were a stroke of genius. In his proof he devised a method of coding statements using what he called Gödel numbers.

After his father's death in 1929, his mother and brother had moved to Vienna. In 1933 Gödel was invited to join the Institute for Advanced Study at Princeton. Even so, Gödel periodically returned to Austria, where he lectured. During this time he suffered emotional crises resulting in a number of nervous breakdowns in Vienna. Were they precipitated by his father's objection to Gödel marrying the nightclub dancer, Adele Prokert Nimbursky? Were they related to his hypochondria? Did he have trouble coping with the consequences of his fame? He did have trouble dealing with everyday social situations. At Princeton he had a small circle of friends — his closest friend being Albert Einstein — and he had no interest in expanding his circle, though many reached out and made overtures to him. He never enjoyed being the focus of attention, and had trouble facing controversy. He went to unusual extremes to avoid people. For example, if someone he did not want to meet, contacted him for an appointment, instead of declining the engagement, he would make the appointment, but never show up. His logic was that this would insure he would not run into that individual since he would not be at the agreed location and time. In 1938 his personal life took a positive turn when he married Adele, the nightclub dancer to whom his father had objected. They remained married his entire life. His yoyoing back and forth between the United States and Austria came to abrupt end in 1939, when on a trip to Austria he was attacked and beaten by fascist students. In addition, when he arrived in Austria, he received an anonymous letter indicating that his associations with Jewish-liberals were known. This immediately motivated Gödel to apply for a visa to the United States, which was quickly granted. When he returned to Princeton, Gödel received a permanent appointment to its Institute for Advanced Study. Yet, it was thirteen years before became a full professor of mathematics. Why was such a renowned mathematician not promoted sooner? Was there uncertainty about how he would fulfill the routine duties of a professorship? When he was appointed, he took his job very seriously, and met its obligations very successfully and thoroughly. Among the many recognitions for his accomplishments were: an honorary degree from Yale in 1951, the Einstein Award, an honorary degree from Harvard in 1952, National Medal of Science in 1975. He refused an honorary degree from Austria. During his later years he turned his attention more to philosophical inquiries, especially in mathematics. Gödel retired in 1976, at the age of 70, having spent over 36 years at Princeton. In his final years he questioned his achievements and the value and impact of his work. Is this what gnawed at him? It did add another layer to his disturbed psyche.

After his death his private papers revealed some interesting idiosyncrasies. Among them were boxes of unanswered letters requesting information, opinions, reactions to the works of others. Here were found drafts of replies, often rewritten a number of times, but never mailed. Were these written to satisfy his own needs? Was he such a perfectionist that it was difficult for him to formulate a response with which he was completely satisfied? Did

To every ω-consistent recursive class κ of *formulae* there correspond recursive *class-signs* r, such that neither υ Gen *r* nor Neg (υ Gen *r*) belongs to Flg (κ) (where υ is the free variable of *r*).

Gödel's Incompleteness Theorem as it appeared in 1931 in his paper "On Formally Undecided Propositions in Principia Mathematica and Related Systems I." In basic terms it says — a formal system of thought deduction will produce at least one true statement which the system cannot prove, thus making the system incomplete.

his paranoia extend to his personal thoughts, never feeling comfortable about discussing or revealing his beliefs? Among his copious notes appear his thoughts about the objectiveness of mathematical ideas. Being a very religious person, he even wrote a proof for the existence of God. To his mother he wrote:

We are of course far from being able to confirm scientifically the theological world picture...What I call the theological world view is the idea, that the world and everything in it has meaning and reason, and in particular a good and indubitable meaning. It follows immediately that our worldly existence, since it has in itself at most a very dubious meaning, can only be the means to the end of another existence. The idea that everything in the world has a meaning is an exact analogue of the principle that everything has a cause, on which rests all science.[5]

Was his mind being flooded with ideas that made everyday mundane circumstances seem futile to deal with? Why did he have difficulty even accepting a casual appointment? What triggered him to go over the edge? Who did he believe was his enemy? Why did he believe someone was out to poison him? Was there a physical condition that exacerbated his mental instability?

The why of Gödel's death remains a mystery.

---

[1] Oskar Morgenstern was a renowned mathematics economist and co-invented *game theory* with John von Neumann.

[2] Königsberg is famous for Euler's solution to the *Königsberg bridge problem* in 1736. His solution launched the field of topology.

[3] An axiomatic system cannot be all encompassing, e.g. the field of arithmetic which is created by establishing axioms and definitions formulating and proving theorems is not capable of proving all true statements. This system is not completely air tight, true statements can be devised which that system cannot prove.

[4] *Life in Gödel's Universe: Maps all the way* by George Zebrowski. *Omni*, April 1992. page 53. Originally from *Mind Tools* by Rudy Rucker. Houghton Mifflin Co., Boston, 1987.

[5] From *Pi in the Sky* by John D. Barrow. Clarendon Press, Oxford. 1992. page 124.

Newton's apple
NEVER WAS

"Thank you, Mrs. Barton, for consenting to meet with me." Voltaire graciously greeted Catherine Barton, Sir Isaac Newton's niece.

"I am delighted to have this opportunity to speak of my esteemed uncle Isaac," she replied.

"He was buried like a king who had done well by his subjects, Mrs. Barton. Words cannot describe his greatness and genius," Voltaire added.

"Over the years your words have meant much to us. My uncle was so special. I have only wonderful things to say about him. Such a dear man. Do you know what my husband John says?" Mrs. Baron asked.

"As a matter of fact, I do not. Do tell me," Voltaire probed.

"Uncle Isaac never desired recognition for his wonderful

Isaac Newton (1642-1727) as a young man.

inventions. He was content to let others have the glory of his work[1]. If it were not for his friends and countrymen, his greatness would have passed unnoticed. He was such a humble man," she said smiling.

"I'll be sure to include this in my writings. But do you have any favorite anecdote you want to share?" Voltaire asked in anticipation.

"The apple story most certainly," she replied.

"The apple story?" he questioned.

"Yes. How the law of gravity dawned on my Uncle Isaac, when an apple fell on his head."

"You mean a falling apple sparked Sir Isaac Newton's ideas

about gravity?" Voltaire asked in amazement.

"Exactly," replied Mrs. Barton.

Myths are part of the human tradition — spanning the centuries from classic to modern times. Newton's apple falls into the same category as George Washington's cherry tree. In Netwon's case, one of his many biographers, Sir David Brewster, perpetuated this myth which was originally passed on to Voltaire by Newton's niece, Catherine Barton. So began the legend of the apple hitting Newton on the head prompting him to discover his law of gravity[2]. Voltaire was such an admirer of Newton, that he can be described as one of Newton's advocates. In fact, today he would be characterized as Newton's French publicist, even though they had never met. Voltaire is one of the many responsible for spreading and perpetuating this *Chicken Little* story about Newton.

Newton's mental feats and fame reached mythological stature — rendering him almost divine. People were beginning to attribute to Newton things that were not his, as Volatire noted:

> *"There are people who think that if we are no longer content with abhorrence of a vacuum, if we know that air has weight, if we use a telescope, it is all due to Newton. Here he is the Hercules of the fable, to whom the ignorant attributed all the deeds of the other heroes."*[3]

The majority of the populace could not intelligently discuss or understand his works. The apple story was something people could comprehend and relate to easily. It was something that could be repeated without an in depth explanation. And so the apple story crossed the English channel. However, not everyone believed it. In fact, mathematician Carl F. Gauss had this to say about it in the 1700s:

> *"Silly! A stupid, officious man asked Newton how he discovered the law of gravitation. Seeing that he had to deal with a child intellect, and wanting to get rid of the bore, Newton answered that an apple fell and hit him on the nose. The man went away fully satisfied and completely enlightened."* [4]

One of many postage stamps depicting Newton's apple.

And so, the fable continues today — a part of mathematical pop culture. Newton and the apple are inseparable.

---

[1] John Conduitt was the husband of Catherine Barton. He succeeded Newton as Master of the Mint, and was in charge of preserving Newton's papers. Yet his statement about Newton humble character cannot be substantiated. On the contrary, Newton was very protective of his works.

[2] Newton, using Galileo's three laws of motion to formulate his law of gravity, which is described by the equation, $F=(Gm_1m_2)/d^2$. If m1 represents the mass of the earth, m2 the mass of the moon, d the distance between the centers of the earth and moon, and G the gravitational constant, then F is the force of the gravitational attraction between the earth and the moon. Newton declared that this formula held between any two bodies of the universe. Newton guessed the value of G, although it was actually determined in 1798 by an experiment performed by Henry Cavendish.

[3] From *Let Newton Be!* by John Fauvel. Oxford Univeristy Press, Oxford. 1989. page 4.

[4] From *Against the Gods* by Peter L. Bernstein. John Wiley & Sons, Inc., New York. 1996. page 139.

MATHEMATICAL
"BROOKLYN BRIDGE"

"But where did you get these letters? They are marvelous!" Michael Chasles asked, very excited about the documents Vrain-Denis Lucas had shown him.

"I have done extensive research, searched in many places and traveled to many distant lands specifically seeking out old and ancient documents and writings. It has been my personal quest in life, much more than a hobby. But I am very reluctant to part with any," Lucas replied.

"Lucas, what use do you have of these mathematical correspondences. I am especially interested in those you have between Pascal and Newton. There is a greater question at stake here. It is my duty to make these known. You must help me set history straight. These letters show Newton does not deserve all the credit and fame for the theory of gravity. Please, I will pay you generously." Chasles was almost pleading.

Lucas hesitated just the right amount of time, and then replied, "I suppose since you put it in those terms, it is my way of contributing something to humanity. In my possession they are but a selfish indulgence. They should be in the hands of historians and museum curators, rather than locked up in my home."

"Yes, yes, exactly! When can you bring them to me," Chasles asked eagerly. "Don't fear, I will take good care of them, make sure they get into the proper hands, and I will pay you dearly. These are precious documents. When will you bring them?"

"Tomorrow. Tomorrow morning," Lucas replied.

Blaise Pascal (1623-1662)

* * *

"These are exquisite. The paper reflects their age. France will show England who was first!" Chasles was blind with excitement as he read over letters exchanged between Pascal and Newton. Amazingly these established the contributions of Pascal to ideas that up until then were considered to have been the sole work of Isaac

Newton. "Do you have other documents?" Chasles asked eagerly.

"Yes, I have many more. But I need to cherish them for a while before parting with them," Lucas said.

"Yes, of course. Take as long as you need, but when you are ready, remember to come to me. Rest assured I will pay you their value," Chasles reassured him.

"I have no doubt of that. I trust you completely," Lucas answered.

* * *

"Why these are incredible. Letters between Pascal and Galileo," Chasles said, hardly believing his eyes.

Alexander the Great

"Those are not the most important. In these boxes you will find letters form Alexander the Great to Aristotle, from Cleopatra to Julius Caesar, even one from Mary Magdalene to Lazarus," Lucas said, holding an enormous box of papers.

"I will take them all, and will take my time going through them and sorting them." Chasles took the box from Lucas and once again gave him an envelope filled with francs.

"If there is anything else you need, you know where to reach me, mon amie." And with that Lucas left.

* * *

If Chasles had opened that box with the so called ancient documents then, he would have found that they were all written on antiqued paper, and all written in French!

How was it this renowned 19th century mathematician so duped by a con man?

Michael Chasles (1793-1880) was a well known French mathematician of the 1800s who specialized in the field of geometry and published a number of works that became quite famous[1]. In addition, he was a professor of mathematics at the École Polytechnique in 1841, and in 1846 the Sorbonne established the chair of geometry for him. The Copley medal of the Royal Society of London was awarded to him in 1865. His credentials were impeccable. That is until he was approached by Vrain-Denis Lucas, con man extraordinaire. He apparently realized Chasles' *Achilles heel* — his patriotism and his interest in mathematical history. Lucas set up an incredible sting, researching dates and people of various periods, and then forging letters that

42

would be of historical interest. He took care to write the letters on paper that he acquired or prepared to look exceptionally authentic. We are not talking about a few documents, but thousands. He painstakingly produced these letters from 1861 to 1870. It proved to be a very lucrative job for him, allegedly preparing over twenty-seven thousand pieces for which Chasles paid over 140,000 francs! What blew Lucas's cover?

Chasles presented the letters, allegedly written between Blaise Pascal and Isaac Newton, to the Academy of Sciences hoping to prove that Newton was not the originator of the law of gravity. At this time it was discovered that the handwriting in Chasles documents did not match that of Pascal's, which the Academy had in its archives.

If Chasles had taken time to look at and question all the different documents when he purchased them — letters from such famous people as Alexander the Great, from Plato, Cleopatra, Mary Magdalene — he assuredly would have wondered why they were all written in French and all written on paper.

Even though Lucas served time in prison for his scam, Chasles suffered from this embarrassing situation, temporarily losing face and credibility among his fellow colleagues. What a blow to the ego of one specifically trained in logic!

---

[1]Among these are *Aperçu historique sur l'origine et dévelopment des méthodes en géometrie* first edition in 1837 on historical development of geometry — *Traité de géométrie supérieure* published in Brussels in 1852 —*Traité des sections coniques* printed in Paris in 1865.

CHRISTIANS MURDER HYPATIA

Synesius[1] —

I looked over your work on the Diophantian problem — it is excellent. I might suggest another approach to the second part, which I have outlined on your enclosed work. See what you can do with the second problem I proposed.

Now, about your words of caution. Thank you, my dear friend, for your concern. I know Cyril[2] is a fanatic, but I cannot imagine he would harm me because of my mathematics, my work as a teacher at the Museum, or even because I would not become a Christian. Orestes[3] has also pleaded with me to leave. He feels Cyril will now focus his anger on neo-Platonist and pagans, having succeeded in forcing the Jews to flee Alexandria. I cannot run away nor remain silent against such injustices. It goes against everything I believe in. I cannot stop doing what is my life and work. I will not conceal or change my beliefs.

Even if my life is in danger, what good is a life that one does not live as one chooses?

I promise I will be careful.

—Hypatia

*   *   *

Hypatia—

Please trust me. You have under estimated Cyril. He is a man driven by intolerance and ignorance. He feels your mathematics is evil and your influence as a teacher over people of various beliefs disturbs him. You have not embraced Christianity, and this will drive him and his followers over the edge. His new position, his new power make him especially dangerous. Heed Orestes' warnings. He will not be able to protect you. I plead with you.

Keep safe.

—Synesius of Cyrene

*   *   *

A warm March day in 415 A.D..

After having engaged her students in a brilliant philosophi-cal discussion, Hypatia guided her chariot confidently down the streets of Alexandria toward her home. Up ahead she noticed a crowd had gathered in front of the Caesarium Church, which she decided it best to avoid. Before she could turn her chariot, two men pulled her down.

"Release me," she demanded. The mob of Christians was very angry. They had been fired up against Hypatia. The instigator yelled, "She has poisoned Orestes' mind against Cyril. Kill the pagan!".[4] They dragged Hypatia into the church. Tearing off her clothes, they brutally killed her by scraping her flesh from her bones with sharp shells. They then cut up her body and took it to Cinaron to burn.

The wheel had come full circle. The persecuted had become the persecutors.

H ypatia's (370-415 A.D.) gruesome death was recorded by 5th century Christian historian, Socrates Scholasticus. But Hypatia's fame does not only rest upon the circumstances of her death. She was the first woman mathematician and philosopher to be widely recognized in history. She was born in Alexandria during turbulent times of power struggles between Romans and militant Christians. Her father, Theon, was a distinguished mathematician and astronomer who taught at the Alexandria Museum, which was associated with the Library of Alexandria and came to be known as the university. He was known for his work on Euclid's *Elements*

and Diophantus' *Arithmetica*. An enlightened man and recognizing his daughter's talents, will, and desire to learn, he took charge of her education even though the times were not favorable nor encouraging for women to be educated. Much has been written about her intelligence and her beauty. Following her father's example, she taught mathematics and philosophy at the Museum. In fact, they worked together on books regarding Euclid's and Diophantus' works. She considered herself a neo-Platonist, a pagan, and a follower of Pythagorean works[5]. In conjunction with her work at the university, she wrote many treatises and books on mathematics for her students. Her work focused on Euclidean geometry and the works of Diophantus, and she authored a popular treatise on the conics of Apollonius. In addition to solving the the traditional Diophantian equations, she developed new solutions and new problems for her students. She was an engaging lecturer, and her mathematics courses were very popular. Interested in mechanics and applied science, she invented various instruments including — an astrolabe, a hygrometer, a water level instrument, and a water distiller. Although none of her works survive, her history has been preserved through extant books and works she coauthored with her father, through the letters of her pupils, and by historical accounts.

As an intellectual and philosopher, she was involved in discussions on such topics as politics, religion, sciences. As her student Hesychius wrote:

> *"Donning the philosopher's cloak, and making her way through the midst of the city, she explained publicly the writings of Plato, or Aristotle, or any other philosopher, to all who wished to hear... The magistrates were wont to consult her first in their administration of the affairs of the city."* [6]

The political and religious unrest of the time were certainly the cause of her death. After reporting her death to Rome, Orestes requested an investigation. The investigation never took place supposedly because of the lack of evidence and witnesses. What transpired is uncertain, but we do know Orestes apparently resigned and left Alexandria. Speculation as to who the murderers were points to the Parabolan monks of the Church of Cyril and the Nitrian monks. Did Cyril order her slaying? Unknown.

Hypatia became a legend in her own right. The nature of her death had a profound negative impact on educational freedoms in the exploration and expression of ideas during the years that followed. Although tremendous progress has been made since the year 415 A.D., the intolerance and ignorance responsible for her death remain.

---

[1] Synesius was a pupil of Hypatia who went on to become a wealthy and powerful Bishop of Ptolemais.

[2] Cyril was Patriach of Constantinople.

[3] Orestes had been a former student of Hypatia and at the time of her death was a friend and the Roman Prefect of Alexandria.

[4] An anti-Roman demonstration did take place. The mob was worked up against pagan philosophies, scientific teachings and Roman sympathizers, and it seems that Hypatia was a well known figure and a perfect scapegoat to demonstrate their will and power.

[5] She lived about 700 years after Pythagoras.

[6] McCabe, Joseph, *Hypatia, Critic,* 43, 1903. page 269.

Cantor
driven to nervous
breakdown

"As you can see Cantor has gone too far. His crazy ideas have led him to a breakdown," Kronecker smiled wryly as he addressed Weierstrass.

"I cannot agree with you," Weierstrass interrupted. "Cantor is a very intense person. Brilliant. Driven by his work. It was not his work that drove him to a nervous breakdown. You know that."

"Ridiculous!" Kronecker countered. "And besides, I do not call his work mathematics."

"That is exactly what I am getting at. Ridicule is at the core of his mental problems," Weierstrass added.

"Look at the crazy ideas that are now surfacing in mathematics. Playing with infinities leads to all sorts of anomalies. Best to ignore such inconsistencies. How can a mathematician consider such ideas mathematics — these monsters and infinite numbers. "

Georg Cantor (1845-1918)

"I disagree," Weierstrass said emphatically. "You should keep an open mind. These ideas have been evolving for many years. It is how mathematics has always blossomed. You cannot ignore or squelch the innovative mind."

The two mathematicians continued their discussion as they had done many times in the past. Kronecker insisted on preserving the classical approach to mathematics, not recognizing the validity of indirect proofs and the need or existence of fractals, transfinite and transcendental numbers. Weierstrass was willing to delve, explore and support the new mathematics that was being presented.

Karl Weierstrass(1815-1897)

The 19th century witnessed the surfacing of extraordinary mathematical ideas — ideas such as the theory of infinite sets, transfinite numbers, non-Euclidean geometries, fractals. With the exploration of the infinite, traditional mathematicians were shaken from their complacency. Many were opposed to introducing or even considering topics dealing with the infinite. Well aware of the paradoxes and problems associated with infinities, they were very content to ignore mathematics that dealt with infinite sets[1].

In 1884 at the age of 40, Georg Cantor, one of the most brilliant, creative, and innovative mathematicians of the 19th century suffered his first in a series of nervous breakdowns. What precipitated these breakdowns remains speculative. Yet historical evidence reveals a very insidious and calculated attack on his work and his mathematical ability as forged by Leopold Kronecker. Kronecker was a successful businessman and a respected and adequate mathematician. He volunteered his teaching services at the University of Berlin, and accepted a professorship at the University when his mentor, Ernst Kummer, retired in 1883. Cantor had studied at various outstanding European schools of mathematics, and in 1867 received his doctorate from the University of Berlin where he had been a student of Kronecker's. Kronecker had substantial influence on university hiring and on the selection and content of articles submitted to journal publications. He used his prominent position to launch subtle attacks against Cantor's mathematics and to muster the backing of other conservative mathematicians. Kronecker seemed to know just how to incense Cantor, while maintaining a low profile for himself.

His attacks seriously impacted Cantor's career. Unable to get a position at the University of Berlin, Cantor settled for the less known University of Halle. During this time of discontent, Cantor had hoped for a public airing of his mathematics. He was confident in his mathematics saying:

> *"My theory stands as firm as a rock; every arrow directed against it will return quickly to its archer. How do I know this? Because I have studied it from all sides for many years; because I have examined all objections which have been made against the infinite numbers; and above all because I have followed its roots, so to speak, to the first infallible cause of all created things."* [2]

Kronecker was too clever to enter into an open debate, but exercised his power by electing not to publish Cantor's articles in the journal he edited. Cantor became increasingly paranoid, but not without reason. Kronecker even wrote to the journal *Acta Mathematica,* which had previously published some of Cantor's work, indicating he was planning to submit an article that would dismiss the usefulness of modern set and function theories. Learning of this, Cantor felt there was a conspiracy between the editors of both journals , and unfortunately did not submit additional articles to *Acta Mathematica,* thereby playing into Kronecker's plan.[3]

What exactly was it that Cantor created/discovered that so railed many conservative mathematicians? He described and defined numbers that until this time had not been known — the transfinite numbers. Springing off of ideas of mathematicians who explored the infinite, such as Zeno, Aristotle, Galileo, Gottfried Leibniz (1646-1716) and especially Bernhard Bolzano(1781-1848) and J.W.R. Dedekind (1831-1916), Cantor brilliantly created an arithmetic that dealt with infinite sets.[4] His theory of infinite sets showed how the set of whole numbers, the counting numbers, the even counting numbers, the odd counting numbers, the rational numbers all had the same (cardinal) number of elements, namely the transfinite number aleph null, $\aleph_0$. He further demonstrated that all infinite sets did not have the same number of elements[5] — that there were different infinite cardinal numbers for these sets. His work relied heavily on the use of indirect proofs, which many traditional mathematicians felt were not rigorous or trustworthy.

Although traditional mathematicians freely used infinite series and the real numbers, they refused to recognize or deal with infinite sets. Cantor realized his work was very controversial, but fortunate-

Leopold Kronecker (1823-1891)

ly he had the conviction and drive to pursue it and to not be inhibited by ridicule.

Infinities had frustrated mathematicians for centuries—

• Galileo, having worked with infinite sets, was unable to determine which of the sets — the whole numbers or the even whole numbers is larger — finally concluded "...infinity and indivisibility are in their very nature incomprehensible to us."[7]

•Carl Gauss insisted "...against the use of infinite magnitudes...which is never permissible in mathematics."[8] Gauss felt the infinite could only be applied to limits of ratios.

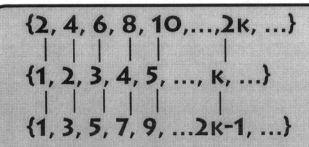

By using the clever technique of paring up elements in sets with the set of counting numbers, Cantor showed that the even and odd counting numbers have the same number of elements as the set of all counting numbers — these three infinite sets have the same number of elements, which he called aleph null, $\aleph_0$.

Kronecker contended "God made integers, all the rest is man made." In an 1885 letter to Sonya Kowalewski, Karl Weierstrass writes "... Kronecker delivers himself the following verdict which I repeat word for word: 'If time and strength are granted me, I myself will show a more rigorous way ... and they will recognize the incorrectness of all those conclusions with which so-called analysis works at present' ... It is sad ... to see a man ... let himself be driven by the well justified feeling of his own worth to utterances whose injurious effect upon others he seems not to perceive."[9]

And indeed Kronecker carried on his vindictiveness against Cantor for over 10 years. Even after Kronecker's death in 1891, his assault on Cantor's mathematics left many mathematicians doubtful and suspicious. Criticized and undermined, Cantor suffered a series of nervous breakdowns which spanned more than 30 years. Fortunately in between bouts he was able to resume his ingenious work. In 1918, he died in a mental institution in Halle. He did receive

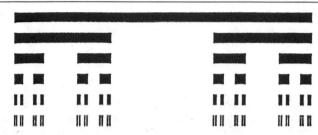

The first stages of the Cantor set constructed by Georg Cantor in 1883. Many mathematicians ridiculed such work and referred to such objects as mathematical monsters. Today they are called fractals, and belong to the study of fractal geometry.

$\aleph_0$ number of elements in each of the following sets—
{1,2,3,4,5,...} {...,–3,–2,–1,0,1,2,3,...} {rational numbers}

$\aleph_1$ number of elements in each of the following sets—
{points on a line}{points in a sphere} {points in a cube}

$\aleph_2$ number of elements in each of the following sets—
{all curves}

$\aleph_3$, $\aleph_4$, $\aleph_5$, ..., $\aleph_n$, ...

Cantor's transfinite numbers and what they measure.

some recognition before his death from some mathematicians and at the 1897 Congress of Mathematicians in Zurich. A beautiful tribute was given him by David Hilbert when he said that Cantor's transfinite numbers were "the most astonishing product of mathematical thought, one of the most beautiful realizations of human activity in the domain of the purely intelligible. ...No one shall expel us from the paradise which Cantor has created for us."[10]

---

1 Fleix Klein, Henri Poincaré, Hermann Weyl, Du Bois-Reymond, Leopold Kronecker were among the mathematicians opposed to the mathematics of transfinite numbers.

2Page 198, *Pi in the Sky* by John D. Barrow. Calrendon Press, Oxford, 1992.

3Kronecker probably had no intention of submitting any article.

4 As he pointed out "All so-called proofs of the impossibility of infinite numbers begin by attributing to the numbers all the properties of finite numbers, whereas the infinite numbers...must constitute quite a new type of number as opposed to the finite numbers, and the nature of this new kind of number is dependent on the nature of things and is an object of investigation, but not of our arbitrariness or our prejudice." From *Contributions to the Founding of The Theory of Transfinite Numbers* by Georg Cantor. Translator and editor, Philip E.B. Jourdain. Dover Publications, Inc., New York, 1955.

5He first defined that two infinite sets have the same transfinite number if their elements can be put into a one-to-one correspondence with each other. Then he showed how, for example, the even counting numbers can be put into a one-to-one correspondence with the whole numbers. Therefore, the cardinal number describing both was the same number, namely $\aleph_0$. Cantor developed an ingenious proof to illustrate how the rational numbers could be put into a one-to-one correspondence with the counting numbers, thereby illustrating they had the same transfinite number, $\aleph_0$, of elements. Using an indirect proof Cantor also proved the number of elements in set of real numbers is a greater transfinite number than $\aleph_0$. Cantor went further, showing that the real numbers which were the union of the rational and irrational could also be described as the union of the algebraic and transcendental numbers. Proving the algebraic set's transfinite number was $\aleph_0$ while the set of transcendentals was larger. He also explained the existence of infinitely many transfinite numbers.

6From his publication in Crelle's *Journal* 1874. The English translation of two of Cantor's papers (1895 and 1897) was published in *Contributions to the Foundation of the Theory of Transfinite numbers*, edited by P.E.B. Jourdain, 1915.

7From the English translation of Galileo's *Two Sciences.* page 18.

8Gauss in a letter of Scumacher July 12, 1831, *Mathematical Thought from Ancient to Modern Times,* vol. 3 by Morris Kline, Oxford University Press, Oxford, 1972.

9Page 200, *Pi in the Sky* by John D. Barrow. Calrendon Press, Oxford, 1992.

10From the article *Sur l'infinit* published in the journal *Acta Mathematica,* 1926.

The mathematician who pleaded insanity

"Year in and year out the Nile floods. Though it has its advantages, the waters cannot be controlled. Is this not true?" Alhazen asked the group at the party.

"You are not telling us anything new, Alhazen," a distinguished looking man replied.

"But what I will now say — none of you has ever heard — none of you has even ever dreamt of."

"Enough of your theatrics. Tell us what you have discovered," the host insisted.

"I can construct a machine... a machine so special that it would be able to regulate the Nile's flooding," Alhazen boasted.

"Preposterous!" shouted the host.

"Now you have gone too far," the distinguished man said.

"It is true," Alhazen insisted. "I promise I can do it."

Little did Alhazen know at the time that his ignorant claim would make him a prisoner in his own home. He made a crucial error that would cost him his freedom for years — he tried to solve a problem without having first analyzed all the facts.

Alhazen[1] (965-1039) is remembered today for his contributions to the field of optics. The earlier years of his life were spent under house arrest after his career took an unfortunate turn. Born in Iraq in the town of Basra, he subsequently moved to Cairo, Egypt. Legend has it that during his first year in Cairo he was so taken by witnessing the annual flooding of the Nile that he felt he could devise a hydraulic system to control the power of this river. He made his outlandish claim without having studied the topography or the source of the Nile. At this time Fatimid caliph al-Hakim was ruler of Egypt. He took the work of scholars and scientists seriously, and was responsible for founding an extensive library in Cairo. On the other hand he did not like anyone to take advantage or make a fool of him, and would not hesitate killing any offenders. Having heard of Alhazen's claim, he probably jumped at the chance to have the Nile tamed under his rule. He immediately commissioned Alhazen, who launched an expedition to explore

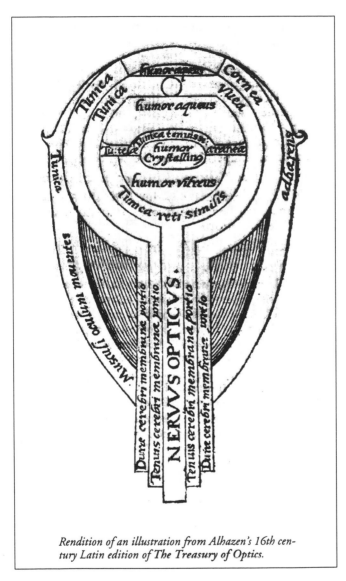

Rendition of an illustration from Alhazen's 16th century Latin edition of The Treasury of Optics.

higher elevations in search of the source of the Nile. The more Alhazen traveled the more he began to realize his "plan" would not be feasible. He returned to Cairo and admitted his gross error. The caliph immediately demoted him. At this point Alhazen became concerned for his life, fearing the caliph might believe he took advantage of his patronage. Alhazen realized he had one option to insure his safety — to pretend he was insane. In those times, the insane were offered special protection. Acting as if he were crazy, Alahzen was placed under house arrest. He kept up his pretense of insanity until 1021 when Hakim died.

When he was not feigning insanity, Alhazen made some phenomenal findings in optics. He rejected the Greeks' idea that light emanated from the eyes, but reanalyzed and expanded upon ancient ideas, especially those of Ptolemy. He described his concept of light as rays and mathematically developed their paths of reflections off of polished surfaces. He even wrote about the connection of the optic nerve with the brain. Studying the eye, he determined how light entered the eye, and worked with the effects of lens. He learned how magnification was due to the curvature of the surface of lens. He constructed parabolic mirrors, and made a pinhole camera. By observing twilight he was able to estimate the depth of the atmosphere. *The Treasury of Optics* is considered his most important treatise. It was translated in the 16th century into Latin and was an important reference for such scientists as Kepler and Descartes.

---

[1]He was known as Alhazen in the Western world, but his real name was Abu'Ali Alhasan ibn-al-Haytham.

THE SCANDALOUS TREATMENT OF ALAN TURING

Thoughts flooded Alan's mind...I can't go to jail. It would mean giving up my work. ... A year in prison would not be physically trying, yet how would I be able to pursue my work there? What would happen to my projects? It would be a mental prison. I would abhor the shackling of my mind more than the indignity of submitting to this "scientific" treatment for my homosexuality.

Turing chose to sacrifice his body rather than his mind for a year. He wrote the following letter to Phillip Hall in April of 1952 describing his choice and situation:

> "...I am both bound for a year and obliged to take this organo-therapy for the same period. It is supposed to reduce sexual urge whilst it goes on, but one is supposed to return to normal when it is over. I hope they're right. The psychiatrists seemed to think it useless to try to do any psychotherapy...."[1]

Alan Turing in essence became a human guinea pig and was subjected to experiments using drugs to reverse his homosexuality. Drug therapy for homosexuality was experimental. As a result of the treatments Turing became impotent and his breasts became enlarged. We can only speculate about the mental and emotional impact of the drugs. At the time scientists believed that the

Alan Turing

side effects were reversible, but it was not certain. According to a medical authority:

> "There is at least a possibility that oestrogen may have a direct pharmacological effect on the central nervous system ... through experiments on rats ... learning can be influenced by sex hormones, and that oestrogen can act as a cerebral depressant in these rodents. While it has yet to be shown that a similar influence

is exerted in humans, there are some indications clinically that performance may be impaired, though more investigation is needed before any conclusion is reached." [2]

Turing survived his year of "therapy", but at what price?

If mathematician, computer theorist and unsung World War II hero, Alan Turing called the police today to investigate a burglary in his house, the fact of his homosexuality would probably have no bearing on police action. But in the year 1952, the climate made him the accused rather than the victim. The Criminal Law Amendment Act of 1885 defined a male homosexual act as a *gross indecency* and was in violation of Section 11 of the 1885 Act.

Turing had called the police to report items missing from his home. At a follow-up investigation, an over zealous homophobic police investigator suspected Turing was a homosexual. He shifted the questioning from the burglary and badgered Turing with questions about his sexual activities. The thief turned out to be connected to a man with whom Turing was involved. Suddenly the theft was no longer the focus of the police investigation.

During World War II, Turing had been instrumental in devising a machine that broke the so called impregnable German *Enigma* machines which enciphered and deciphered the secret German military codes. His work is credited with helping to shorten the war, and for his contribution he was awarded the OBE (Order of the British Empire). In 1945 Turing envisioned a computing machine, called the Universal Turing Machine, which could be adapted to perform the task of any human computer provided there was enough tape[3] and time to give instructions. Until then the term "computer" had referred to a person performing calculations. In post war years he pioneered electronic computing. In 1954, Turing wrote a paper on the feasibility of building an electronic form of the Turing Machine, which is now called the digital computer.

Turing had always been somewhat of a misfit. Few people, other than his associates, could understand his work. The confidential and classified nature of what he was working on was not allowed to be discussed. Focusing so intensely on his work, his lifestyle was rarely questioned and if it was it was accepted by his colleagues. At Cambridge and Manchester Universities Turing found refuge. Being an isolated individual, he was oblivious to the outside world, and had a false sense of security. He was forthright to the point of naivete and unapologetically homosexual.

The Cold War and the McCarthy era were in full swing. Paranoia and fear of Communist infiltration at all government levels were at their peak in the United States and abroad. The government which had encouraged and supported his invaluable work in cryptanalysis and computers during World War II was now the same government that was fearful of the knowledge he carried. The government, feeling he was a vulnerable security risk because he was a homosexual, ultimately withdrew its support. Max Newman, renowned Cambridge mathematician and professor, testified on

Turning's behalf at his trial. Describing him as "particularly honest and truthful. He is completely absorbed in his work, and is one of the most profound and original mathematical minds of his generation." [4] Asked if he would invite him into his home, Newman replied that Alan was often invited to his home and was a personal friend of both himself and his wife.

Turing's death in 1954 at the age of 42 was a shock to his colleagues, friends and family. There was no indication that he was on the brink of committing suicide. Granted he had suffered much from the unfortunate incident of his trial, but that had been two years earlier. And his drug therapy had ended a year ago. He was found lying neatly in bed by his housekeeper. The post-mortem indicated cyanide poisoning. Potassium-cyanide was found in his house along with a cyanide solution in a jar. By his bed was an apple from which a few bites had been taken. No note, no explanation. A verdict of suicide was given, even though the apple had not been tested. Had the drug therapy taken its toll emotionally? Is there anything to the whispered innuendos that he may have been murdered? That the intelligence service had wanted him dead?

---

[1] *Alan Turning —the Enigma* by Andrew Hodges. Simon Schuster, NY. 1983. page 473.

[2] *Alan Turing — the Enigma* by Andrew Hodges. Simon Schuster, NY. 1983. page 474.

[3] In 1955 Max Newmann pointed out "It is difficult today to realize how bold an innovation it was to introduce talk about paper tapes and patterns punched in them into the discussions of the foundations of mathematics." Page 125 *A Computer Perspective* by Charles and Ray Eames. Massachusetts, 1990.

[4] *Alan Turing — the Enigma* by Andrew Hodges. Simon Schuster, NY. 1983. page 472.

FOURIER COOKS HIS OWN GOOSE

"Don't open that window! " Fourier yelled at his friend as he entered the room.

"But it is boiling in here, Jean. I can hardly breathe. How can you stand it? And besides how can you wear all those layers of clothing while it is so hot outside. Good lord! You even have a fire in the fireplace. What's going on here?" his friend asked, completely dumbfounded.

"You know I have made extensive studies about the properties of heat, and I am convinced that heat has amazing healing abilities. The warmth is soothing to one's weary bones. I am just carrying my theory to steps beyond," Fourier replied.

"But Jean this is not healthy. Your heart cannot take this heat."

* * *

Ｈow did Fourier develop this strange habit?

Jean Baptiste Fourier (1768-1830) wore many hats — a soldier during the French Revolution and under Napoleon, an assistant lecturer at the École Polytechnique, and a mathematician/ scientist who worked on explaining the proper- ties of heat. His fame

Jean Baptiste Joseph Fourier (1768-1830)

was cinched by a series of mathematical ideas that were derived by unintentionally making a number of errors that would lead him to formulating ideas and a theorem which would take mathematicians over 150 years to correctly justify. Despite these "errors", in 1812 the Academy of Sciences in France awarded Fourier its Grand Prize, not for accuracy of his work, which was fraught with flaws in reasoning, but for his general conclusion. A conclusion which had eluded other famous and very talented mathematicians[1]. Fourier, in his work on describing the mathematics behind the theory of heat, had concluded that *any function* or graph could be described by a series of trigonometric functions. Today, Fourier series and the Fourier integral are studied by calculus students and used by math- ematicians. His work on waves had far reaching ramifications. *Théorie analytique de la chaleur (The Analytical Theory of Heat)*, his major work, was completed in 1822. Fourier formulated his theory

Gaspard Monge (1746-1818) was instrumental in the development of analytical geometry.

In 1798 he and mathematician Gaspard Monge accompanied Napoleon on his campaign to Egypt. Here, Fourier as secretary of Institute d'Egypte, became politically involved in Napoleon's negotiations and diplomatic affairs. It was here that Fourier was introduced to Egypt's desert and its warmth, and began to pursue his research on heat conduction. While in this warm climate he began to believe in the

while trying to solve problems of how heat flows between points of an object. The properties of heat had long interested Fourier. While studying the complex factors involved in heat conduction he developed the Fourier Theorem[2]. This theorm led to his discovery of the mathematics of waves which included his work in the Fourier series and integral. His fascination with heat played a key role in both the life and death of Fourier.

Napoleon Bonaparte (1769-1821)

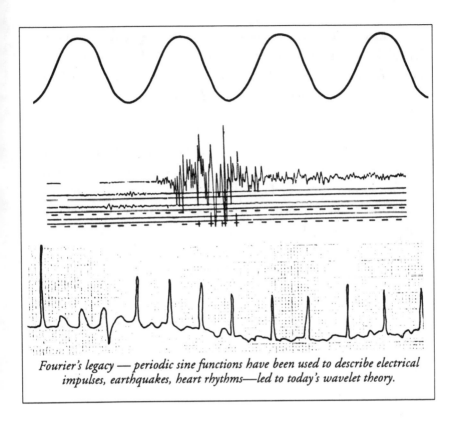

Fourier's legacy — periodic sine functions have been used to describe electrical impulses, earthquakes, heart rhythms—led to today's wavelet theory.

healing powers of heat. Some historical accounts claim he became ill in Egypt with an affliction[3] which heat helped relieve, but accounts vary.

After Fourier returned to France in 1801, he continued his scientific inquiries about heat. He was so captivated by the powers of heat that he probably felt that in matters relating to health, the more heat that was applied the greater the benefit. He carried this belief to the extreme, keeping his quarters unbearably hot and covering himself with layer upon layer of clothing. Undoubtedly this exacerbated his heart condition. He ventured out of his home less and less. Some accounts have him dying from heart failure, others from a fall down a flight of stairs. Perhaps he suffered a heart attack while descending his stairs. He died twelve days after his fall.

Today Fourier is famous for his mathematical analysis of wave phenomena. Be they sound, light, water, earth, or other waves, Fourier series are applicable to all. His work still acts as a stepping stone for the development of new ideas in wave theory such as in Fourier windows and wavelets.

---

1 When Fourier developed his formula for coefficients in the trigonometric series, he was unaware that Leonhard Euler had already developed it. But unlike Fourier, Euler believed it only applied to a small class of functions. In addition, Fourier used sines and cosines to study heat flow similar to the way Daniel Bernoulli used them for his work on vibrations. Other mathematicians who worked on these problems, but failed to see what Fourier noticed, were Joseph Lois Lagrange and J. d'Alembert.

2 Fourier's Theorem states that any periodic oscillation can be described by the sum of simple periodic waves (simple trigonometric functions).

3 One account labeled his condition as myxedema (hypothyroidism) in which the thyroid gland shuts down body functions making one feel the cold more acutely. There is no definitive proof, however, that he had this condition.

The secret work of Carl Gauss

"I cannot understand how a man as intelligent as Archimedes failed to invent a placevalue system," Gauss proclaimed to a friend one day.

"Oh, he probably was too busy with other work to be concerned about it. I am sure he found computation boring, and chose to focus on other ideas," the friend replied.

"But just think how advanced science would now be if only Archimedes had made that discovery!" Gauss said emphatically.

"True, but maybe it was something he felt he did not need to know or use. It just did not interest him. But, you should talk, Gauss. You rarely publish any of your fantastic discoveries. Just think of all the ideas that could be spinning off from them if you would just put them out there

Carl Gauss (1777-1855)

for other mathematicians to think about. You write things in your scientific diary, but rarely share your gems. Why do you keep so much of your work secret?" his friend confronted him.

Gauss was momentarily taken aback by his friend's candor, and then replied, "I study and discover things for myself."

"Well, in that case, don't criticize Archimedes," his friend replied.

What was the real reason Gauss kept so much of his work secret?

It seems that for whatever reasons, Gauss did not practice what he advocated for Archimedes. Had he been willing to share his discoveries, mathematical advances in different areas might have occurred more quickly or in different directions. Why was he so secretive or protective of so much of his work? Of the many outstanding things he discovered[1], he chose to make public only a few.

Between the years 1816 and 1818 the French Academy of Sciences offered a prize to the first person who proved or disproved Fermat's Last Theorem. Gauss had been urged to enter the competition, but his reply was "I am quite obliged for your report of the Paris prize. But I must say that Fermat's theorem, considered as an isolated proposition, interests me very little; I could very easily propose a whole string of such propositions, which no one would be able to prove or use."[2] Imagine all the mathematicians who have expended thousands of hours trying to prove Fermat's Last Theorem! They could have focused their time on other problems, if Gauss had solved it.

At the age of nineteen Gauss had one of his discoveries published. It appeared in the journal *Intelligenzblatt der allgemeinen Literatur-zeitung* in June of 1796. His paper described how he constructed a regular 17-sided polygon[3] using only a straightedge and compass. What was especially significant about this polygon is that it had eluded the ancient Greeks. With this publication Gauss immediately became known in mathematical circles. It was also at this time that Gauss began his *Notizenjournal* — his mathematical/science diary. His first entry was his discovery about constructing regular polygons. He had 145 subsequent entries. The entry dates indicate several of his discoveries predate by many years the same discoveries made by other mathematicians. Among these were the generalization of the double periodicity of elliptic functions and the discovery of hyperbolic geometry. Why didn't Gauss publish these? Why did he withhold his work?

Gauss himself contends that he worked solely for himself and for self-knowledge. He was fortunate his benefactor, the Duke of Brunswick, took care of his expenses. From 1791 until the Duke's death in 1806, Gauss did not have to concern himself with earning a living — all he had to do was concentrate on making mathematical discoveries. When he did publish a piece of work, he was ever so careful to be certain it was a finished piece — clear, precise and perfect. Among his published works were the famous Fundamental Theorem of Algebra[4] (his doctoral thesis) and the Fundamental Theorem of Arithmetic[5]. After the Duke died, Gauss was appointed director of the Göttingen Observatory of Germany. Here too he had almost complete freedom to pursue work that intrigued him. As a result, his studies, discoveries, and inventions went uninterrupted. His diary substantiates that many of these predated those published by other mathematicians. Again we ask, why did Gauss withhold his work? Was he fearful of criticism? Was he too much of a perfectionist? One story tells how he felt humiliated when his work *Disquisitions arithmeticae* was rejected by the French Academy of Sciences. It was then he supposedly decided not to publish anything unless it was perfectly polished. Interestingly, when the Academy of Sciences did a thorough search of its archives in 1935, they found that Gauss had never submitted that work. Was Gauss so thin skinned that he was fearful of making a mistake or receiving any kind of criticism? Gauss may not have wanted to expend the needed effort and time to make his findings

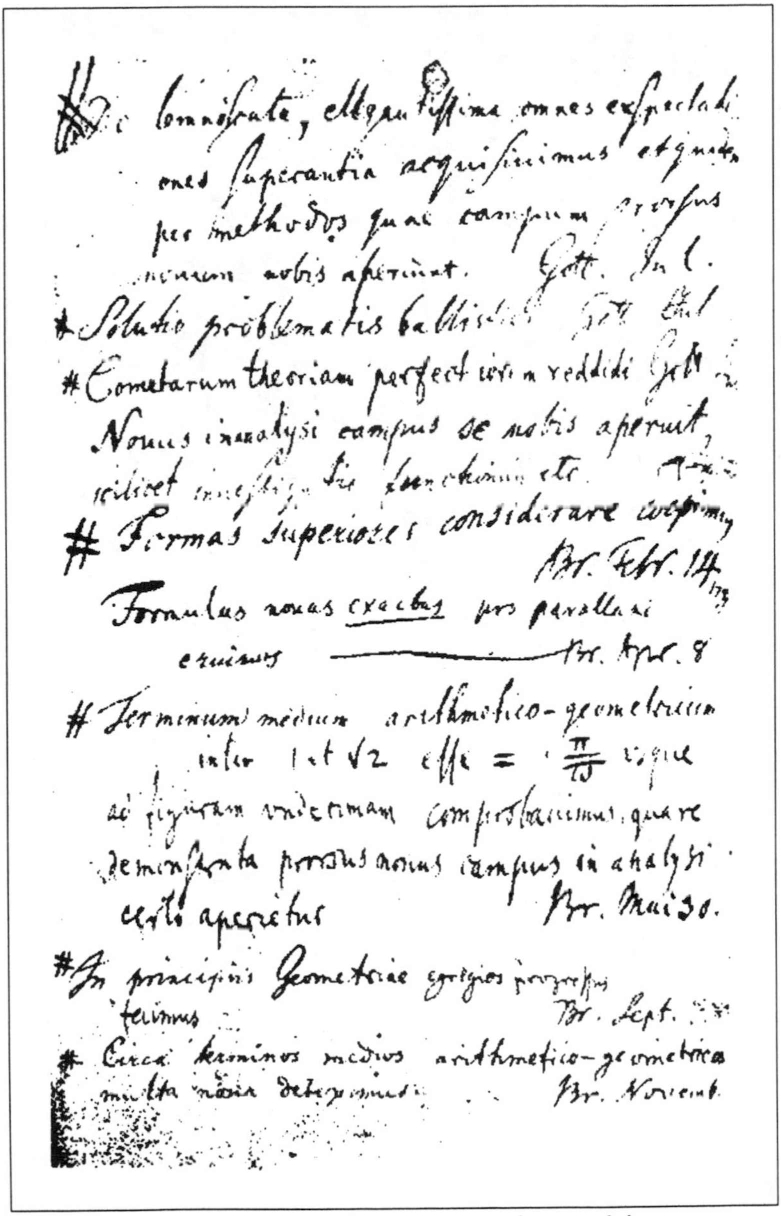

…e lemniscata, elegantissima omnes exspectationes superantia acquisivimus et quidem per methodos quae campum prorsus novum nobis aperiunt. Gott. Jul.

Solutio problematis ballistici Gott. Jul.

Cometarum theoriam perfectiorem reddidi Gott.

Novus in analysi campus se nobis aperuit, scilicet investigatio functionum etc.

Formas superiores considerare coepimus Br. Febr. 14

Formulas novas exactas pro parallaxi eruimus Br. Apr. 8

Terminum medium arithmetico-geometricum inter 1 et $\sqrt{2}$ esse $= \frac{\pi}{\varpi}$ usque ad figuram undecimam comprobavimus, qua re demonstrata prorsus novus campus in analysi certo aperietur Br. Mai 30.

In principiis Geometriae egregios progressus fecimus Br. Sept.

Circa terminos medios arithmetico-geometricos multa nova deteximus Br. Novemb.

A rendition of a page from Gauss' mathematical diary

July 10, 1796

$$\mathrm{E}\upsilon\rho\eta\kappa\alpha$$
$$num = \Delta + \Delta + \Delta$$

October 11, 1796

Vicimus GEGAN

April 8, 1799

REV. GALEN

- *The entry for July 10, 1796 was written in Greek with Archimedes famous declaration — Eureka (I discovered). The mathematical equation he wrote means that every natural is the sum of three triangular numbers. (A triangular number belongs to the set {1,3,6,...,(1/2)n(n+1),...} in which every number represented by dots forms the figure of a triangle, for example 6 gives* ⠿

- *Of the 146 entries in his journal, the meanings for October 11, 1796 and for April 8, 1799 still remain a mystery.*

ready for publication, but chose instead to focus his time and energies on research. Perhaps, for example, he felt his work on hyperbolic geometry was so revolutionary when compared to Euclidean geometry, he feared possible repercussions would tarnish his reputation. His concern is reflected in a letter to a friend:

"...to work up my very extensive researches for publication, and perhaps they will never appear in my lifetime, for I fear the howl of the Boeotians if I speak my opinion out loud." [6]

Regardless of the reason, the self-imposed secrecy of this remarkable genius delayed mathematical advances.

---

[1] Among his bountiful discoveries were—work on electricity, geodesy, complex numbers, the theory of functions, convergence of series, number theory, the Fundamental Theorem of Arithmetic, the Fundamental Theorem of Algebra, modular arithmetic, proving the law of quadratic reciprocity, solutions to problems of determining planetary orbits, principles of hyperbolic geometry, his methods of least squares, and the list goes on.

[2] *Carl Friedrick Gauss* by Tord Hall. MIT Press, Cambridge MA. 1970. page 151.

[3] Gauss' discovery was farther reaching than constructing the 17-sided polygon. He showed that regular polygons with a prime number, p, of sides can be constructed using a straightedge and compass if and only if p is of the form $2^{2n}+1$. Up until then regular polygons with 3, 4, 5, 6, 10, and 15 sides had been constructed by the ancient Greeks. Using Gauss' formula, $p=2^{2n}+1$, n= 0 and n=1 gives the polygons with 3 and 5 sides, with which the Greeks had dealt. Note the formula does not give the prime number 7, thereby showing that a 7-sided regular polygon cannot be constructed with only a compass and a straightedge. But when n=2, then p=17 a prime number which according to Gauss' work is constructible with the compass and straightedge, which Gauss did perform. In addition, his work showed that any prime number resulting from his formula would be a constructible regular polygon.

[4] This theorem states that every algebraic equation has a root of the form a+bi (a complex number). He further showed how these numbers could be graphed on a plane.

[5] This theorem shows that every natural number can be represented in one and only one way as the product of primes.

[6] *Journey Through Genius* by William Durham. John Wiley & Sons, Inc. NY, 1990. page 55.

FEMALE MATHEMATICIAN CRASHES THE old boys' club

"Sophie, what are you doing awake so late?" Sophie's mother asked as she entered the candlelit room.

"I'm just reading, mother."

"Reading what?" her mother asked.

"Oh, books from father's library," Sophie responded, trying to avoid giving a specific answer.

"Let me look," her mother said as she picked up a book from the bed.

"Euclid's **Elements.** Mathematics books! That's what you're reading! Haven't your father and I forbidden you to study mathematics? You know it is not good for your health or your mind."

"How could a few ideas harm my health? I so enjoy learning

about these things, especially since I must stay inside most of the day because of all the political unrest."

"We've told you we do not want you learning mathematics. These ideas should not trouble the head of a young girl — this is man's business. Since you have not heeded our wishes, we will have to take extreme measures. You may not have a fire in your room at night, and we will lock up your clothes. In addition, I am taking your candles away from you. That should prevent you from wandering to the library and staying up late. No more reading. No more mathematics. Is that clear, Sophie?" her mother demanded.

"It is clear," Sophie replied. Under her breath she said, "It still won't stop me from studying mathematics."

This was one of the many obstacles Sophie Germain encountered in her quest for mathematical knowledge.

*   *   *

"I have a paper here written by M. LeBlanc," Professor Lagrange said addressing his class at the École Polytechnique. "It is an exceptionally well done and original piece of work. Will Monsieur LeBlanc come to my office after class so I may personally discuss and commend him on his fine work." With that Lagrange dismissed the class.

Outside the lecture hall, Robert, one of Sophie's friends from Lagrange's class, rushed up to her. "Did you hear, Sophie?"

"Hear what?" Sophie asked.

"Professor Lagrange wants to congratulate LeBlanc for his outstanding paper. He wants to see you in his office."

"He wants to see LeBlanc in his office?" Sophie asked. "What will he say when he finds out Monsier LeBlanc is a mademoiselle?" Sophie wondered.

S ophie Germain's determination and thirst for mathematical ideas were the driving forces which helped her seek her goals. Circumventing her parents' restrictions, she had her secret stash of candles, and at night, to keep warm she would wrap herself in blankets and go off to her father's library for books. Her desire to learn was greater than her parents' obstinacy. Eventually, they gave in. The French Revolution and the Reign of Terror isolated and confined Sophie to her home, but fortunately it also afforded her the opportunity to relish her father's books. She became her own instructor in mathematics.

Having exhausted her father's library, Sophie needed to look elsewhere for information. The École Polytechnique of Paris, where many of the foremost French mathematicians lectured, had recently been established, but there was a problem — women were not allowed to attend. Did this stop Sophie? No! She selected courses that interested her, and studied borrowed lecture notes from

Sophie Germain (1776-1831) alias M. Le Blanc

friends. She must have been especially enthralled by Jospeh Lagrange's course on *analysis* because she decided to submit a paper on work she had done in connection with the course. Naturally she had to submit it under a male pseudonym — M. LeBlanc. Even though Lagrange was surprised to discover that LeBlanc was Sophie Germain, he did not discriminate against her because she was a woman. Instead, he encouraged her and praised her work. In addition, he took it upon himself to introduce Germain to many French mathematicians and scientists. Although she still could not attend a formal school, she continued her studies by correspondence with the many contacts she had now made.

In 1801 Carl Gauss published his *Disquisitiones arithmeticae,* which dealt with number theory. Germain was able to get a copy, and it captivated her. As a spin-off from Gauss's work, Germain derived some ideas of her own which she wanted to share with Gauss. Never having been introduced to him, she decided she would write him, and once again use the pseudonym M. LeBlanc. Gauss was impressed with LeBlanc's work, and began a lengthy correspondence with LeBlanc. Germain would not have been exposed if it were not for her concern for Gauss's welfare during the French military campaign in Germany. She asked the French general to send an emissary to be certain he was safe at his home near Breslau. Gauss was thoroughly confused when the emissary brought up Germain's name. She explained her alias in a subsequent letter in which she wrote:

> *"...I have previously taken the name of M. LeBlanc in communicating to you those notes that, no doubt, do not deserve the indulgence with which you have responded ... I hope the information that I have today confided to you will not deprive me of the honour you have accorded me under a borrowed name, and that you will devote a few minutes to write me news of yourself."* [1]

Gauss replied as follows:

> *"The taste for the abstract sciences in general and, above all, for the mysteries of numbers, is very rare: this is not surprising, since the charms of this sublime science in all their beauty reveal themselves only to those who have the courage to fathom them. But when a woman, because of her sex, our customs, and prejudices, encounters infinitely more obstacles than men in familiarizing herself with*

84

their knotty problems, yet overcomes these fetters and penetrates that which is most hidden, she doubtless has the most noble courage, extraordinary talent, and superior genius."[2]

Many doors were closed to Germain simply because she was a woman, yet her desire to learn and do mathematics persevered. In 1816 she was awarded the French Academy of Science Grand Prize for her work on analysis of vibration in plastic surfaces. In 1831, Gauss recommended that Germain receive an honorary doctorate from Göttingen University. Unfortunately, she succumbed to the cancer she had been battling for two years, and she died at the age of fifty-five before the degree could be conferred.

---

[1]From *The History of Mathematics: A Reader,* by John Fauvel and Jeremy Gray. The Open University, London. 1987. p. 497.

[2] ibid 1.

Newton was no sweet cookie

"Come in," Edmond Halley called out in answer to the knock at the door. He was busily at work preparing Newton's **Philosophiae naturalis principia mathematica**[1] for publication.

"Good day, Mr. Halley," Robert Hooke said as he entered.

"Good day, Mr. Hooke. How nice to see you. Please sit down. What is the nature of your visit?" asked Halley.

"As you know Mr. Newton and I have been corresponding about my work on light and gravity," Hooke said. "Since you are in the process of publishing Newton's **Principia,** I thought it only fitting that some acknowledgement be afforded me for my work which I shared with him in this area."

"That seems only fair to me," Halley replied. "I will mention it to Newton."

Sir Isaac Newton (1642-1727)

* * *

"Absolutely not," Newton said emphatically. "I'd rather not print Book III of **Principia** than give him any credit." Newton was in a rage.

"But all Hooke is asking is that you simply acknowledge the

PHILOSOPHIÆ

N A T U R A L I S

PRINCIPIA

MATHEMATICA.

A U C T O R E

ISAACO NEWTONO,

EQUITE AURATO.

EDITIO ULTIMA

Cui accedit ANALYSIS per Quantitatum SERIES, FLUXIONES & DIFFEREN-
TIAS cum enumeratione LINEARUM TERTII ORDINIS.

AMSTÆLODAMI,

SUMPTIBUS SOCIETATIS.

M. D. CCXXIII.

Title page for Philosophiae naturalis principia mathematica

help his ideas may have given you when you corresponded in 1679. We know he would not have followed through or been able to develop those ideas as you did," Halley explained.

"No, no, no!" Newton yelled.

Robert Hooke (1635-1703)

When he would get in these rages, Halley knew it was best to leave him alone. It was not the first time Halley had seen him like this, and undoubtedly it would not be the last. He knew when it came to protecting his ideas or accepting any form of criticism Newton lost control.

* * *

Newton finally let Halley acknowledge Hooke but only after promising to delete the word "Clarissimus" after Hooke's name.

* * *

There is considerable speculation about Isaac Newton's unexplainable mood swings and erratic behavior. What was at the root of his strange and unpredictable behavior?

_____

Edmond Halley (1656-1742) is most known today for describing the elliptic orbit of Halley's comet and succesfully predicting its return in 1758.

W hen one mentions the name Isaac Newton (1642-1727) the word genius comes to mind. Here was a man whose youth and early college years were not especially noteworthy, but who, in the matter of a few years(1665-1666), had formulated ideas on gravity and laws of motion, light, and calculus which would have a profound impact on mathematics and science. The amount of concentration he expended must have been formidable. He would approach a problem by constantly thinking about it until he had solved it.   During these periods of intense effort he disliked distractions, regardless how small.   Add to this some of his other qualities — such as his childlike outbursts, reactions to criticism, and his over protective attitude of his works — and one had a very volatile personality.   Not a humble man, Newton was fiercely protective of his ideas, inventions and works. He had difficulty acknowledging the contributions of others, although he is often credited with saying "If I have seen further than other men, it is because I stood on shoulders of giants."[2]   In Newton's time this declaration had almost become a cliche. In fact, the first written record of it dates back to the 12th century. It was the expected response, appearing numerous times in various forms and is even represented in the windows of the Chartres Cathedral.

In the late 1600s, Newton was on the verge of a nervous breakdown.   In fact, he went over the edge a number of times. When he had a mental collapse he would withdraw and become reclusive.   If he was contacted he would usually respond in an irascible manner.

In 1683 his friend Edmond Halley drew him out of his seclusion and coaxed him to prepare his work *Principia* for publication by the Royal Society of London. The Royal Society of London had just experienced financial difficulties with an unsuccessful publication. In addition, the society did not want to risk becoming embroiled in any possible controversies over authorship[3] of ideas that Newton's work might cause. Halley so believed in Newton's work that he chose to underwrite the cost himself. He not only paid for the publication, but also took charge of all the details — getting illustrations, proofing copy, having Hooke acknowledged — all the while appeasing Newton. Halley was able to get the first edition of *Principia* published in 1687. Instant fame followed. Along with the fame came the close scrutiny of Newton's ideas. Newton had always had trouble dealing with criticism, and now was no exception. In 1693 he suffered a serious nervous breakdown. The pressures of work, fame, plus other factors have been put forth as possible causes for his mental crises. Among these are:

— In 1692 his laboratory caught fire and his work and notes were burned. This in itself would be a severe blow to anyone, especially since the luxury of photocopy machines and computer back-up copies did not exist.

— A few months before his breakdown his relationship with Swiss mathematician Nicolas Fatio de Duillier ended. Newton had known Fatio since 1689, but the

exact nature of the relationship is undocumented. Were they lovers or just very close friends? This is unknown, though his relationship with Fatio has been described as "the closest Newton appears to have come in his life to having a warm human relationship (apart from his mother)".4

Nicolas Fatio de Duillier (1664-1753)

We do know that Newton did not have any involvement with women during his adult life. His breakdown occurred in the fall of 1693, at which time he wrote this strange letter to his friend John Locke:

> *Sr*
> *Being of the opinion you endeavored to embroil me with women & other means I was affected with it as that when one told me you were sickly & would not live   I answered twere better if you were dead.   I desire you forgive me this uncharitableness...*[5]

— Was his nervous breakdown the result of mercury poisoning? Besides carrying on his scientific experiments, Newton also immersed himself in the study of alchemy which necessitated his handling and burning mercury. Many scholars have likened his emotional behavior to those actions often induced from mercury poisoning.

"The hazards of tasting mercury compounds and breathing the fumes of sublimed salts are self-evident. No wonder Newton began to have problems. In fact this was about the time when Newton began to suffer the poor digestion and insomnia he spoke of in his explanation of the odd letters of autumn 1693. It may be Newton himself identified the cause as the quicksilver vapours warned of by alchemical writers for centuries before Newton."[6]

After Newton's mental condition stabilized, he did not shy away from entering public life. He moved to London in 1696 to become

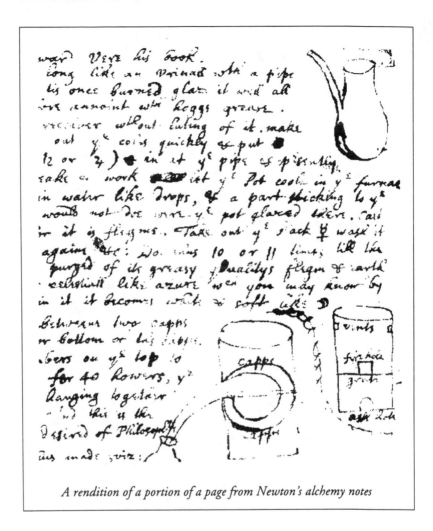

A rendition of a portion of a page from Newton's alchemy notes

Warden of the Mint, and lived there until his death in 1727. In 1703 he was elected President of the Royal Society, and was knighted by Queen Anne. Regardless of his public profile he still obsessively carried grudges, particularly against Robert Hooke, and Gottfied Leibniz (with regard to calculus) and John Flamsteed (with regard to the control of astronomical data). Operating behind the scenes, Newton was most adept at organizing and getting placement for his followers. For example, he was able to get William Whiston the Lucasian Chair at Cambridge, Edmond Halley the Chair of Geometry at Oxford, David Gregory the Savilian Chair of Astronomy at Oxford, and he also saw to it his candidates were given preferential consideration for appointment to lectureships in mathematics at Christ's Hospital. In addition, he was able to get his people appointed secretaries and demonstrators of experiments to the Royal Society. This ingenious network served him well for the rest of his life. Indeed, he was an active partner in the creation of his own mystique.

---

[1] Most often referred to as *Principia*, 1687. Here Newton developed his law of gravity, formulated his laws of motion, and worked out a system for describing the motions of celestial bodies.

[2] This comment by Newton appeared in a letter to Robert Hooke in 1676.

[3] Robert Hooke, a member of the Royal Society, was contending that Newton formulation of the inverse law of gravity was indeed his work.

[4] Gertsen, Derek, *Let Newton Be!*. Oxford University Press, Oxford. 1989. page 19.

[5] Gertsen, Derek, *Let Newton Be!*. Oxford University Press, Oxford. 1989. page 17.

[6] From the *Notes & Records* of the Royal Society of London vol. 34. No. 1 , July 1979. *Mercury Poisoning: A Probable Cause of Isaac Newton's Physical and Mental Ills* by L. W. Johnson and M. L. Wolbarsht.

Where's the Nobel prize for mathematics?

Alfred Nobel sat in his study working on finalizing the details of his will, his mind going over how he had set up the prizes.

"There. Finished. Have I overlooked anything?

"Small gifts to friends and relatives. I mustn't overdo. Large inherited wealth only breeds complacency.

"Executors are to convert all property to cash, and invest it in safe securities. The interest from these investments will pay for the annual prizes to those persons who have contributed most materially to benefit mankind. Yes, that is the way I want it.

"No prize for mathematics. I have not stated that outright, but it will be implied by its omission. I have specified the exact fields that should be awarded. That should suffice. It will be obvious I do not want a prize in mathematics. My instructions will thereby keep Mittag-

Leffler from receiving a prize."

* * *

Why was Alfred Nobel so against the wealthy mathematician Gösta Mittag-Leffler possibly receiving a Nobel prize? What did Nobel have against Mittag-Leffler? Or did Nobel have something against mathematics?

Alfred Nobel

The Nobel Prize for Physics and Chemistry.

Thus far four people have each won two Nobel Prizes— Madame Curie in physics and chemistry, Linus Pauling in chemistry and peace, John Bardeen both in physics, Frederick Sanger in chemistry and physics.

W hen Alfred Nobel died in 1896, a fund of $9,200,00 was set up for annual Nobel Prizes in the areas of Peace, Literature, Physics, Chemistry, and Philosophy.[1]

There has been much speculation about why Nobel did not establish a prize for mathematics. The stories cover a wide spectrum. Some can be immediately dismissed. A mathematician had an affair with Nobel's wife. But we know Nobel was never married. Could it have been that a mathematician had an affair with someone with whom Nobel was involved? This sounds plausible, but there is no evidence to substantiate this story.

Nobel was a shy man, who disliked publicity and was often self-deprecating. At the age of 43, he placed a personal ad in a Vienna newspaper which read "A very wealthy, cultured, elderly gentleman living in Paris desires to find a lady also of mature years, familiar with languages, as a secretary and manager of his household." The ad was answered by an attractive 33 year old, cultured Austrian aristocrat, Bertha Kinsky, who needed employment. They met in Paris, and she took the job. But it was soon evident to Bertha that perhaps Nobel had hoped for more than just a household manager, especially after he asked her if she were "fancy free"[2]. After working only one week Bertha eloped with the lover she had left in Vienna. Nevertheless, she and Nobel remained life long friends. Shortly thereafter, he met Sofie Hess, a 20 year old woman working in an Austrian flower shop. The two had a Pygmalion-like relationship. Nobel first established an apartment for her in Vienna, then one in Paris, and lastly a villa in Germany. Although he wanted her to evolve into a cultured woman, she only enjoyed indulging her expensive tastes. Nobel admonished her in letters for her frivolous actions and use of money, but she did not take heed. Even though Nobel repeatedly tried to end the affair, she would continually write for more money. Even when she later became pregnant and married the father of her child, Nobel did not cut off her support. Apparently none of the men in the lives of the women with whom Nobel had contact were mathematicians.

So what caused Nobel's animosity toward mathematics? Did he have a falling out with a mathematician? There are many references to the mathematician Gösta Mittag-Leffler with relation to Nobel. Some references insist that Nobel definitely did not want the Swedish mathematician to receive his prize. Did Nobel have a business problem with Mittag-Leffler? Did Nobel not approve of some of Mittag- Leffler's business dealings? Had they been friends, and later had a falling out? One stipulation in Nobel's will for literature was that the prize be given to "whomever, within literature, has produced the most outstanding work, in an ideal direction." When his will was published, many wondered how this statement should be interpreted. Mittag-Leffler came forth and contended that Nobel "meant anything that takes a polemic or critical standpoint toward Religion, the Monarchy, Marriage, the Establishment as a whole."[3] Was he implying that Nobel was an anarchist? Had they argued over these issues?

Nobel was a very adept businessman, but if it were not for his inventiveness, he might not have been so successful. Nitroglycerine had been developed in 1847 by the Italian Ascanio Sobrero. They worked together in the same laboratory, but Sobrero felt that this substance was too dangerous to consider using it commercially. Obviously Nobel did not. He envisioned many uses for this explosive compound — building tunnels, railways, clearing roads, mining, weapons of war. Before releasing it commercially, he developed the detonating cap for exploding it. Even so, this lethal substance took its toll. In 1864, Nobel's younger brother and four other factory workers were killed in a factory explosion in Stockholm. This incident began to label Nobel as a manufacturer of destruction. Sweden refused to let him rebuild his laboratory. Similar tragedies occurred in other factories around the world. He was not deterred, however, but was determined to find safer means to harness the power of nitroglycerine. To minimize danger to others, he decided to conduct his experiments on a barge. It was here he accidentally happened upon an important discovery. One of the containers on the barge had leaked nitroglycerine into the packing material of diatomaceous earth. When Nobel discovered the container's problem, he realized that the nitroglycerine was safer to handle when contained in this earth-like material. In fact, it could not be detonated without a detonating cap — which led to his invention of dynamite. In subsequent years he invented blasting gelatin, smokeless powder — labeled by the public as vehicles of war. This did not seem to deter Nobel's work. In fact he felt his inventions would help end war, stating to his dear friend Bertha von Sutter (formerly Bertha Kinsky) "my factories may make an end of war sooner than your congresses ... because the day that two armies have the capacity to annihilate each other within a few seconds, it is then likely that all civilized nations will turn their backs on warfare."[4]

Bertha von Sutter at this time was very active in the European peace movement. It was her work and influence, coupled with negative public opinion for his inventions, that influenced Nobel in establishing the Nobel Prize for Peace. In fact, the first Nobel Prize for Peace was awarded to Bertha von Sutter.

And so, although the mystery of why there is no prize for mathematics is unresolved, we laud Nobel for his recognition of contributions in so many other fields, only wishing mathematics had also been included.

From Left to right —
Nobel Prize for Physiology & Medicine, Nobel Prize for Literature.

The Nobel Prize for Peace.

---

[1]In recent years the Swedish government established the Nobel Prize for Economics, which is subsidized by the government and not by Nobel's fund.

[2] *While he expected the worst, Nobel hoped for the best,* by Donald dale Jackson. *The Smithsonian,* November 1988. page 201.

[3] *The Literary Nobel Prize* by Kjell Espmark. Published by G.K. Hall & Company, Boston. 1991.

[4] *While he expected the worst, Nobel hoped for the best,* by Donald dale Jackson. *The Smithsonian,* November 1988. page 218.

Was Galois jinxed?

As the young man lay on the ground, blood seeping from his wound, his mind was flooded with equations and ideas.

The symbols and numbers suddenly disappeared, as he felt a hand on his shoulder and a voice asking, "Are you alive?" A passerby had nearly stumbled over his limp body. He had been laying there several hours since suffering a gunshot wound from one of the frivolous duels that were the vogue of the times.

"Yes," the young man gasped, barely able to reply.

"I am going to take you to a doctor," the voice said optimistically.

"Will I live?" the young man of only 20 years wondered. As he felt himself being carried away, the ideas returned to his mind. Ideas he had not had time to fully develop during his frantic night of writing and creating . But now his time had run out. His hectic night of creativity had been filled

Évariste Galois (1811-1832)

writing and analyzing his mathematical revelations. Thoughts he so desperately wanted to share. His anxiety was fueled by his belief that the end of his life was near.

"Will they understand what I was trying to show?" Galois wondered. "If I only had more time. More time to elaborate, write details of the ideas I outlined.

Galois died in 1832 at the age of 20—his innovative contributions to mathematics unrecognized. What went awry in the life of this young man? A genius — a child prodigy. Misfortunes upon misfortunes plagued his short life. Can genius be considered bad luck? In the case of Évariste Galois it certainly was. Imagine a mind brimming with incredible mathematical ideas, innovative techniques and solutions with only limited avenues open to share them. He was too advanced for his time. Not many of his contemporaries were capable of understanding his insights — his approach was so original and modern, and his explanations too sketchy and brief. Many had trouble filling in the steps he saw as obvious. It took over 14 years after his death for some of the essential parts of his work on group theory and algebraic equations to finally become available to mathematicians.

Galois and his family lived in the French village of Bourg-la-Reine. His father, a liberal republican, was the mayor of the village. During his early years, his mother, an educated but eccentric woman, took personal charge of his education. When he was twelve his parents decided to enroll him in a formal boarding school in Paris. Classics were the major emphasis of the school, and he initially excelled because his mother had focused on them. When he was introduced to mathematics, the subject captivated him. In no time his mathematics instruction and textbooks were too elementary. He sought out sources, devouring the geometry of Legendre and the algebra of Lagrange. Capable of doing computations and problems easily and in his head, he often intimidated some his teachers, who insisted on seeing every step of his work. Some misinterpreted his attitude, thinking him defiant and insolent. Becoming increasingly disenchanted with his other studies, he did not apply himself in these areas. Some teachers even insisted he be demoted. H. J.

Vernier, one of the first teachers to become aware of Galois's special talent and unusual drive, tried to encourage him to learn mathematics in a systematic fashion. But Galois was too impatient. Not challenged by the elementary work, he wanted to go directly to advanced mathematics. He had no trouble understanding the mathematics, but had trouble communicating his ideas and findings to people who needed to see the complete picture of what he was trying to say. Like many young people, he became cocky. At the age of 17, against his teacher's advice, he decided to take the entrance exam to the École Polytechnique without having prepared properly. He failed, and blamed his failure on his examiners and the system. Continuing at his school, he took more advanced courses taught by an exceptional teacher, Louis Paul Emile Richard, who immediately recognized Galois's brilliance and tried to foster his genius. During this time Galois showed his superiority in mathematics, and concentrated more and more on his own ideas rather than on his classwork. In April of 1929, Galois's first paper, *Proof of a Theorem on Periodic Continued Fractions,* was published in *Annales de Gergonne.* Having worked intensely on the theory of equations, in May of 1829 Galois submitted a paper containing his fundamental discoveries to the Academy of Sciences which appointed Augustin Cauchy as referee of Galois's paper. Cauchy promised to present Galois's work to the Academy at its next session. Having such a prominent mathematician present one's work would guarantee the attention of the Academy. Unfortunately, Cauchy did not keep his promise. Cauchy was expected to present both Galois's paper and one of his own, but he claimed he was unable to attend that session of the Academy and asked to be rescheduled for the next session. At the next session in January, Cauchy only presented his own work.[1] Needless to say, this situation disillusioned Galois and reinforced his negative feelings about academia.

Augustin Cauchy (1789-1857)

In July of 1829 Galois's father was the victim of a vicious slander perpetrated by a young priest who wrote, signed, and publicly circulated a vicious verse with the mayor's name. The father was mortified and committed suicide near the school Galois attended in Paris. Villagers were so incensed by the clergyman's dirty tactics against the popular mayor, that his father's funeral was marred by a riot.

Galois absorbed himself in mathematics. Desiring to attend the École Polytechnique, the foremost French school of top mathematicians and scientists, he decided in August of 1829 to take the entrance exam again. But this exam was also to be ill fated. During the oral part of his test the examiner challenged a result involving logarithmic series. Rather than explaining his work, Galois insulted the examiner's intelligence by saying the result was obviously clear. He failed his exam once again. His only option now was to take the entrance exam for the teachers' school, École Préparatoire. His outstanding mathematics grade gained him entrance at the end of 1829. Feverishly continuing his research in mathematics, he wrote three papers developing revolutionary ideas in mathematics, including work on the theory of algebraic equations. He submitted this new work, *On the condition that an equation be solvable by radicals*, to the Academy of Sciences in February of 1830 for the Grand Prize[2] in the mathematics competition. This original work probably would have earned him the prize, but once more a stroke of bad luck interfered. This time his paper was safely in the hands of the Secretary of the Academy, Joseph Fourier, who decided to review it at home. Fourier died in April 1830 before he could examine it, and it was never found among his things. Galois was thoroughly embittered saying, "Genius is condemned by a malicious social organization to an eternal denial of justice in favor of fawning mediocrity."[3]

1830 was a turbulent time in France. Desiring to take a stand and become involved in the revolution, Galois wrote a letter to *Gazette des Écoles* in which he criticized the director and students at his college for their political apathy. He was expelled.

No longer in school, he decided to teach his own class with materials he prepared on algebra, but unfortunately he could not hold onto his students. At this point he decided to enlist in the artillery unit of the National Guard. While in the artillery, he was encouraged by mathematician Simeon-Denis Poisson to submit his work on the general solution of equations to the Academy of Sciences. He did so in January 1931. Meanwhile, Galois continued his involvement in the Republican cause. He was arrested, and charged with being a traitor and disloyal Louis Philippe (King of France from 1830 to 1848). Even though he was acquitted he was now labeled a radical republican. For his second arrest he was sentenced to six months in jail for the minor charge of wearing an artillery uniform which had been declared illegal by Louis-Philippe. In prison, his only salvation was his work on mathematics. It was here he learned his paper had been rejected. Poisson stated "his argument is neither sufficiently clear nor sufficiently developed to allow us to judge its rigour".[4] Galois was despondent.

Shortly after his release he was taunted into a duel over a young woman with whom he had become involved. Convinced he would die in the duel, he spent the entire night feverishly working on his mathematical ideas and discoveries. Before his death he entrusted his work and his other papers to his friend Auguste Chevalier, asking him to have them read by mathematicians Jacobi or Gauss.[5] Although his work was published in *Revue Encyclopedia* in 1832, it did not impact the development of mathematics at this time. Perhaps it was too obscure and sketchy or not seen by people who understood its potential. Not until 1846 were some of his works presented in edited form in Joseph Liouville's *Journal de Mathématiques pure et appliquées.* By now work on group theory had advanced sufficiently so his discoveries could be appreciated. During his lifetime Galois never experienced recognition for his extraordinary work and advanced ideas, but his legacy has impacted 20th century mathematics.

[1] The Galois legend has been evolving over years. Initially some writers/historians took liberties to embellish the known facts, which ended up being perpetuated. For example, various scenarios have been conjectured as to why Cauchy did not present Galois's paper. Did he suggest to Galois that he combine his work and resubmit it? Had Cauchy lost the paper? Had Cauchy just forgotten to bring Galois's work to the Academy? All these possibilities add fuel to the Galois legend.

[2] In June of 1830 the Academy awarded the Prize jointly to Neil Abel (posthumously) and Carl Jacobi. As a result of their work, Galois wrote three articles in the *Bulletin de Férussac* on elliptic functions and abelian integrals.

[3] Bell, E.T., *Men of Mathematics.* Simon & Schuster, New York. 1965. page 341. It remains part of the Galois lore. It is uncertain whether this is an actual quote or was embellished by Bell.

[4] http://www-groups.dcs.st-and.ac.uk/~history/Mathematicss/Galois.html

[5] Even though Chevalier and Galois's brother sent his mathematical papers to Gauss, Jacobi, and other mathematicians, there is no record of their responses.

I sleep,
THEREfORE I think

"What is that on the ceiling?" Descartes wondered as he came out of a deep sleep. His eyes focused on the fly traveling from one place in the room to another. He was turning over something in his mind, but he couldn't quite grasp the thought.

Ever since he had been a little boy, he had loved to lie in bed and think. He was never rushed to get up because of his frail constitution. Naturally he milked the situation. He so enjoyed this habit. In the quiet of his room his thoughts could flow without interruption. But today the fly had intruded, constantly interrupting his thoughts.

"That's it!" he shouted to himself. Looking at the corner of the ceiling, he saw three imaginary lines (and three imaginary planes) intersecting in one point — the corner. Now his mind raced ...three perpendicular lines meeting in

René Descartes (1596-1650)

one point...imagine each with equally spaced natural number labels ... the corner being zero... every place the fly would alight three numbers could describe its location. It's beautiful ! It's elegant! It's so simple! ... What if there were just two perpendicular lines?" his mind began to think in a new direction. "Then every point in the plane determined by those two lines could be easily described by two numbers. In both cases...in space or in a plane ... an

I sleep, therefore I think

ordered triplet or pair of numbers could be used to identify the location of the point. It is brilliant!" he thought as a big smile came across his face.

* * *

And so the Cartesian coordinate system was born.

A good story, but is it true?

René Descartes (1596-1650) was born into a wealthy family. He no doubt was pampered both because of his frail health and because his mother had died shortly after he was born. He was allowed to stay in bed as long as he wished. Descartes grew accustomed to remaining in the comfort of his warm bed for hours, often thinking or working on things. This habit remained with him his entire life. His father wanted his son to have as many opportunities as possible. When he was eight years old he entered the Jesuit school in La Flèche. After eight years with the Jesuits, he was at a turning point of his life, and decided to go the Paris. There he studied law at the University of Poitier, and received his degree in 1616. Because he was financially independent, he did not need to practice law, but decided upon the military — first in the Dutch army at Breda and later the Bavarian army. Although he engaged in combat, as a gentleman soldier he had ample time to explore his philosophical and mathematical ideas, which seemed to mainly come to him while in bed. This is where and how he conceived the idea of analytical geometry and the Cartesian[1] coordinate system. It took 18 years before these ideas were actually put into print. He remained in the military only a few years before resigning, but continued to travel for several years in Europe before returning to Paris in 1625. In 1628 he decided to settle in Holland where he felt the atmosphere was more receptive and tolerant to his viewpoints and philosophical ideas. He spent the next twenty years of his life in Holland, where he wrote his various theses. From 1629 until 1633 he worked on his philosophy and prepared *Le Monde*[2] for publication. He abruptly decided not to publish it after learning how Galileo and his work had been condemned. At this point he focused on publishing his work on the scientific method (called *Discours de la méthode pour bien conduire sa raison et chercher la vérité dans les sciences)* which included three appendices — *La Dioptrique, Lés Météores,* and *La Géométrie*[3]. In *La Géométrie,* which consisted of about 100 pages, there appeared his first printed notions of analytic geometry. The idea of analytic geometry had also been conceived by Fermat at about the same time, but it was Descartes who first put his ideas into print. When *Discours de la méthode* was printed 1637, Descartes became a celebrity through most of Europe. The Catholic Church did not approve of his works, however, and placed them on their index of prohibited books. This included all his works, even those dealing with mathematics.

In 1649 Descartes made what proved to be a fatal decision. For three years Queen Christina of Sweden had been after Descartes to join her court and become her personal tutor in philosophy. He had no wish to go to Sweden, especially because of the harsh climate; but she was persistent, and he was probably seduced by the

Queeen Christina in her youth

idea of becoming a courtier and by the prospect of life amongst royalty. When Descartes finally accepted, he startled the Queen by choosing to come immediately rather than waiting for the harsh winter to pass. The Queen sent a ship to pick up the famous philosopher. He arrived in Stockholm at the beginning of winter, and was received with much fanfare. His quarters were at the home of the French Ambassador, but little did the 50 year old Descartes know that he would no longer be able loll in bed in the mornings. During his first months in Sweden he realized his student's enthusiasm for philosophy had shifted to the more frivolous interests of the court. Nonetheless Queen Christina eventually sought out the lessons of the philosopher, but on her terms. He was awakened three times a week at 5 o'clock in the morning to give philosophy lessons to the 23 year old energetic and demanding Queen. This early rising in the extremely cold climate reeked havoc with Descartes health. Within a few months he contracted pneumonia, and died within ten days on February 11, 1650. His body was not immediately sent to his homeland. Instead, it remained in Sweden for seventeen years. In 1667 his bones were returned to Paris — or were they? One story relates that his head did not accompany the body. His skull was not returned to France until 1809 when Swedish chemist Jöns Berzelius presented it to French anatomist George Cuvier.[4] Another story contends that his body was sent minus the bones of his right hand. The bones were later procured by the French Treasurer-General.[5] Descartes now rests — intact — at the Pantheon in Paris. His ideas continue to travel in the minds of philosophers and mathematicians.

---

[1] Descartes signed his name in his treatises in Latin, namely Renatus Cartesius— hence the origin of the name Cartesian.

[2] *Le Monde* was not published until after his death in 1664 in Paris.

[3] It was divided into three parts — *part 1* shows the relationship between the fundamentals operations of arithmetic and geometry — *part 2* deals with the classification of curves and methods for finding tangents and normals — *part 3* considers roots of equations and Descartes rule of signs.

[4] From Isaac Asimov's *Biographical Encyclopedia of Science and Technology.* p. 106 Doubleday & Co., Inc,. Garden City, NY. 1972.

[5] From *Mathematical Circles Adieu* by Howard Eves. p.15, Prindle, Weber & Schmidt, Inc., Boston, MA. 1977.

The feud over who invented calculus

"How dare he publish a book on calculus!" Newton fumed to his friend Fatio de Duillier[1].

"The scoundrel stole your work," Fatio said, fueling Newton's anger.

"That's right. He stole my ideas when he was in England as an ambassador."

"We can't let this pass unnoticed. We know about your ideas on calculus. It must not be believed that these were Leibniz's ideas. We will muster the support of your friends. We will show Leibniz. He will wish he had never published his book.

"He will be sorry. This will not pass easily," were Newton's final words on the subject.

Gottfreid Wilhelm Leibniz (1646-1716)

As Herman Hankel said *"In most sciences one generation tears down what another has built and what one has established another undoes. Only in mathematics does each generation add a new story to the old structure."*

The field of calculus is no exception. Its roots originate in ancient Greece. Here the ideas involving infinity were first formally discussed and explored. Here we find Zeno and his paradoxes of motion. Others include Leucippus, Democritus, Aristotle. The Pythagoreans, Eudoxus and Euclid provided the important tool of ratio for measuring rates of change. And here the ideas of Hippias and Dinostratus came into play. It was then Archimedes introduced the use of limits to derive the area formula for a circle. In the centuries that followed the works of other mathematicians from all over the world further explored the ideas that would lead mathematicians to develop calculus. In the 1600s Cavalieri worked with an early form of integral calculus. In more recent centuries among European mathematicians we find Pierre de Fermat, René Descartes, James Gregory, Isaac Barrow, John Wallis. All contributed small yet important steps toward the development of calculus.

*   *   *

In the 17th century two minds in different countries puzzled over problems of change, tangents, maxima, minima, and infinitesimals and formally launched the field of calculus. Gottfried Wilhelm Leibniz in Germany and Sir Issac Newton in England developed their calculus independent of one another. It wasn't the first time that the same mathematical idea had been discovered by different individuals simultaneously[2]. The issue of who discovered what when had not been an issue between the two mathematicians until each had begun to make a name for himself and each began to realize the importance of his discoveries. The dispute ran along national lines. On one side the British claimed calculus was Newton's invention while the Germans' claimed it was Leibniz's.

The situation heated up when Leibniz published his system for calculus in 1684.[3] Leibniz wrote *"My new calculus,...offers truth by a kind of anaylsis and without any effort of imagination — which often succeeds only by accident; and it gives us all the advantages over Archimdes that Vieta and Descartes have given us over Apollonius."*[4] Newton was dismayed that he had not had a chance to publish his version first. Various of his original ideas had circulated, but were not formally published. Newton was infuriated. Friends and fans entered the battle, some claiming Leibniz's visit to London in 1673 had afforded him the opportunity to get his hands on some of Newton's ideas from manuscripts that were being circulated. Did Leibniz see any manuscripts? Shortly before Leibniz died, he wrote in a letter that mathematical correspondent John Collins had shown him some papers Newton had sent him, but maintained these were not of value to him.

Newton's close friend Fatio de Duillier implied that Leibniz had plagiarized part of Newton's work. Leibniz repudiated these accusations. The hostilities continued for years. In 1711 Leibniz decided to appeal to the Royal Society in England about the claim of priority. Yet, even here, things were not viewed impartially as Leibniz had hoped. Leibniz was not allowed to testify on his own behalf. Through the years Newton fueled the dispute from behind the scenes. In addition, other conflicts of interest existed that were not brought in the open. As acting president of the Society, Newton appointed the committee which was to review the situation, and he was careful to chose members that would reflect his interests. The committee's "impartial report" was written in Newton's favor. To add scandal to scandal, the author of the report was not a matter of public record, for Newton himself had written most of the report. Newton did not stop with this. An

Sir Isaac Newton (1642-1727)

anonymous summary of the report appeared in *Philosophical Transactions*, which was also written by Newton. The Latin version of this was aimed at readers on the continent and also appeared in *Commercium epistolicum*.

Newton carried the grudge even after Leibniz had died. In a later edition of Newton's famous work *Principia*[5], Newton altered a small section to be certain that Leibniz would not share any credit. This editorial change was done twelve years after Leibniz's death!

In the meantime, the study of calculus was splintered. The works of these two men were not integrated and refined. The British remained faithful to Newton's version, while the Continent favored Leibniz, whose notation and some other aspects were more advantageous. This schism lasted over 100 years, and in essence cut off British mathematicians from mathematical advances in this area taking place on the Continent.

Today the consensus is that both men developed their ideas of calculus independent of one another. When the cloud of this unfortunate dispute settled, the contributions of both to the development of calculus were appreciated. There was enough glory for both egos. The main loser was calculus — cooperation and discussions would have enhanced rather than hindered its evolution.

---

[1]Nicholas Fatio de Duillier was a young Swiss mathematician, whom Newton had known since 1689. de Duillier's treatise was the primary spark of the calculus feud.

[2]For example, hyperbolic geometry was developed independently by János Bolyai & Nikolai I. Lobashevcky.

[3]Leibniz's book was titled *Nova methodus pro maximis et minimis* (*A New Method for Determining Maxima and Minima*).

[4]*Great Ideas of Modern Mathematics* by Jagjit Singh. Dover Publications, Inc., New York 1959. page 27.

[5]*Principia's* complete title is *Philisophiae naturalis principia mathematica*.

The truth about Einstein & Maric — It's all relative

"Mileva, what do you think of the idea that light is really made up of particles?" Einstein asked his then classmate and lover Mileva Maric.

"It has possibilities. Have you also considered the energy that light particles may carry?" she asked enthusiastically.

"No. ... but of course. That energy must be related to the frequency of radiation. But how?" Einstein wondered out loud. "By Planck's constant. that's how. Don't you agree?" Albert once more asked Mileva's opinion.

" I was wondering when you would arrive at that. I totally agree. The radiation must be equal to the product of Planck's constant and the frequency of the radiation. Is that not what you had in mind?" Now Mileva was asking Albert.

"Exactly! It is such a simple equation. Energy from radiation = Planck's constant•frequency of the radiation. It's beautiful." Einstein declared.

Mileva was always intrigued by how Albert's mind worked. She was happy she could keep up with him, and sometimes even contribute ideas. Lately she was pleased to bring to his attention what she knew about the Michelson-Morley experiment which measured the speed of light in different directions. She knew it would be of help on the ideas concerning relativity that they were discussing. When they had these exchanges the room was electric.

"Have you considered what you would do with regard to the ether ?" Mileva asked the ever gnawing question.

"That is something we still need to work on. Should we go against the established ideas? I am not certain, but anything is possible," Einstein replied. "We will have to think this through. But I do know it is always very exciting to work together."

*   *   *

Could such a dialogue have taken place? It may have been one of many exchanges the couple had over their years together. Was Mileva Maric only Einstein's sounding board or much more? Were they really a team? In recent years, these questions have cropped up. Here was a woman entering the same course of studies as Einstein in 1896. She must have had an incredible amount of determination, desire and interest, since that was not an easy road for a woman in that period. Few had been allowed to attend the universities, and even fewer were awarded advanced degrees. If scientists were skeptical of some of Einstein's early revolutionary ideas, would

Mileva Maric

they have even considered them if presented by a woman? If, indeed, these ideas were Maric's, did the couple decide to be practical and submit them only under Einstein's name? Or, did they even consider submitting them under both names? The answers to these questions vary, depending on who is answering them. What ifs are among the most frustrating forms of speculations. But none the less what if ....

In recent years the life of Albert Einstein — the 20th century intellectual icon — has been subjected to the same exhaustive scrutiny which plagues today's superstars. What kind of man, lover, father was Einstein? Answers to these questions would not change the brilliant ideas attributed to him. But now, even the authorship of his ideas has been questioned with such queries as — Did Einstein develop his theories by himself? —What role did his first wife play in the formulation of relativity theory? —Did they contribute equally? —Was Maric the brains behind the man? All these speculations will probably never be resolved to everyone's satisfaction. On what grounds is such skepticism founded? In 1969 a Yugoslavian publisher (Bagdala) issued *In the Shadow of Albert Einstein* by Desanka Trbuhovi'c-Gjuric. A subsequent German edition was published in 1988 with the author's name now given as Senta Troemel-Ploetz. Incorporated in this edition were letters[1] which had been in the possession of Einstein's eldest son, and had remained in Hans Albert Einstein's safe deposit box until his death in 1986. Among these were letters between Albert Einsetin and Mileva Maric, which both shed some light and cast some shadows on the Einstein saga.

Maric and Einstein met in 1896 as entering students at the acclaimed Swiss polytechnic school, the Federal Institute of Technology[2]. She was twenty-one years old, Einstein was seventeen. At the university they shared notes, exchanged ideas, helped one another get through their course work — normal things college students do. Both families frowned on the relationship. Maric's father was a Yugoslavian government official and her mother was from a wealthy family. It was during these times[3] apart that the couple corresponded. Their letters[4] began as innocent notes.

Albert Einstein

Although their first letters were formal, they intensified as their relationship developed — ultimately reflecting a full blown romance. Parts of Einstein's letters to Maric triggered speculation about Maric's role in his work. The letters[5] also revealed how both changed intellectually and emotionally during these seven years.

Along with his words of affection, Einstein would write comments about physics — including ideas, projects, experiments, questions — to Maric, who, on the other hand, rarely alluded to physics in her letters. His letters make reference to *our* work, *our* theories, *our* investigations. For example, in a letter to Maric in 1901 he writes: "How happy and proud I shall be when together *we* shall conclude victoriously *our* work on relative motion."[6] In another letter he says: "I'm so lucky to have found you a creature who is my equal, and who is as strong and independent as I am! I feel alone with everyone except you."[7] Were these *ours* and *we* just figures of speech? Just words of a lover wanting to share and include his lover in his work? There is no definitive answer.

In 1900 Einstein and Maric both took final exams in order to graduate and qualify as instructors. Maric did not pass. At this point she had two options — submit a thesis for her PhD or retake the exam. Einstein passed his exam, thus making him eligible to teach, but apparently his professors did not recommend him for a university position at this time. Consequently he was only able to find tutoring jobs and temporary work as a teacher. Maric scheduled to retake her exam the next year, but she too found herself in a predicament — she was pregnant and unmarried. Her emotional state must have been fragile, making preparation for her exam difficult. Einstein kept in contact with her, trying to be encouraging. He would write her about his experiments but would add such comments as "How is it going with our little son and your paper for the PhD."[8] Frustrated over his difficulty in finding a permanent job, Einstein wrote her the following: "I have decided the following about our future. I shall look for a position, however poor...As soon as I have found one I will marry you... Although our situation is very difficult I am quite confident since having made this decision.[9]"

In light of the circumstances, it was not surprising that she failed her exam when she retook it in late 1901. Not wanting to be alone, she returned to her parents' home in Yugoslavia, giving birth in January of 1902 to a baby girl she named Lierserl. Although Einstein writes positively about their daughter, for reasons unknown, they did not choose to raise their child. It is not known what happened to Lierserl, but she was probably given up for adoption or raised by a member of Maric's family.

All during this time Einstein parents had been pressuring him to get out of the relationship. His mother thoroughly disapproved of

Maric, and Maric's parents were not thrilled with Einstein. In June of 1902 Einstein landed a junior position as a clerk with the Swiss Patent Office in Berne. The two lovers followed their own minds, and married in a civil ceremony on January 6, 1903. In the summer of 1903 they set up their home in an apartment in Berne. In 1905 Einstein published his three most important papers in *Annalen der Physik* — the first dealt with his hypothesis about molecules randomly distributed in fluid[10] — the second talked about the nature of light[11] — the third on the special theory of relativity. Once the impact of these ideas began to be understood, Einstein's fame grew. He remained at the Patent Office until 1909 when he received an academic appointment at the University of Zurich. From then on his career blossomed. His relationship with Maric became strained, and Einstein eventually asked for a divorce which was finalized in 1914. Among the terms of the divorce were the following stipulations: — Maric be responsible for the care and education of their sons[12] — Einstein be responsible for child support — 40,000 marks were to be deposited in a Swiss bank, its interest at Maric's disposal — if Einstein were to receive[13] a Nobel Prize, the prize money was to be given to Maric.

Besides Einstein's alluding in his letters to their working together, what other evidence of Maric's contributions can be found? The Einstein skeptics contend that — Maric worked with physicist Paul Habricht on the development of a machine to measure small electric currents which was patented under the name of Einstein-Habricht[14] — the couple may have agreed to keep Maric's contribution secret to enhance Einstein's chance for a university appointment[15] — Russian physicist Abraham Joffe (who was an assistant to an editorial board member of the *Annalen der Physik* when Einstein submitted his articles) was supposed to have seen

Einstein's three famous original papers of 1905 signed with the name Einstein-Maric. Unfortunately Joffe is deceased, and the original no longer exists. What does remain from Joffe is a short memorial he published shortly after Einstein died in which he says, " In 1905 there appeared three articles in *Annalen der Physik* ... the author of these — unknown at the time — was the clerk at the Patent Bureau, Einstein-Marity."[16]

Is there sufficient evidence to discredit Einstein? It is circumstantial, and less powerful when one considers it in the light that Mileva Maric herself never discredited Einstein. In later years, even though Maric was bitter enough to feel like lashing out at Einstein, she never did[17]. Did she feel no one would believe her? Or was it because she knew Einstein was the one who deserved credit? Yet another Einstein mystery! Perhaps as Einstein would say, "It's relative."

---

[1]Forty-one of these letters are by Einstein to Maric, and only eleven are from Maric to Einstein.

[2]Marcs was initially going to study medicine at the University of Zurich, but changed her major to mathematics and physics.

[3]From 1896 until they were married in 1903, Albert and Mileva's relationship bloomed even though very close family commitments necessitated they spend periods of time apart (school holidays, summer breaks, vacations.)

[4]Not all these letters have yet been printed.

[5]There are among fifty-four letters published in the book, *Albert Einstein/ Mileva Maric, The Love Letters* (by J. Renn and R. Schulmann. Princeton University Press, Princeton, N.J.. 1992. Not all their letters have been printed.

[6]Page 9 from *Einstein Lived Here* by Abraham Pais. Clarendon Press, Oxford, 1994.

[7]Page 8 from *Einstein Lived Here* by Abraham Pais. Clarendon Press, Oxford, 1994.

[8]Einstein had assumed the child would be a boy, by making reference to their son. Page 9 from *Einstein Lived Here* by Abraham Pais. Clarendon Press, Oxford, 1994.

[9] Page 9 from *Einstein Lived Here* by Abraham Pais. Clarendon Press, Oxford, 1994.

[10]This paper dealt with Brownian motion.

[11]He put forth here that the energy of light is carried by individual units called quanta, which went against traditional physics which had been around for a century. Needless to say his ideas on the nature of light were received with skepticism and only experimentally verified ten years later by American physicist Robert Andrews Millikan.

[12]Their first son, Hans Albert was born in 1904 and Eduard was born in 1910.

[13]Einstein received the Nobel Prize in Physics in 1921, at which time his award, approximately $32,000, went to Maric.

[14]*Shadow of Albert Einstein* by Senta Troemel-Ploetz. German edition, 1988.

[15]*Shadow of Albert Einstein* by Senta Troemel-Ploetz. German edition, 1988.

[16]Page 15-16 from *Einstein Lived Here* by Abraham Pais. Clarendon Press, Oxford, 1994.Schulmann. Princeton University Press, Princeton, NJ. 1992.

[17]She only complained that he did not help enough with the care of their younger son, who had been institutionalized. She probably influenced her elder son's embittered feelings toward Einstein, though the two men seem to have reconciled prior to Einstein's death.

HIERONYMI CAR

DANI, PRÆSTANTISSIMI MATHE

MATICI, PHILOSOPHI, AC MEDICI,

ARTIS MAGNÆ,

SIVE DE REGVLIS ALGEBRAICIS,

Lib. unus. Qui & totius operis de Arithmetica, quod

OPVS PERFECTVM

inscripsit, est in ordine Decimus.

HAbes in hoc libro, studiose Lector, Regulas Algebraicas (Itali, de la Cossa uocant) nouis adinuentionibus, ac demonstrationibus ab Authore ita locupletatas, ut pro pauculis antea uulgo tritis, iam septuaginta euaserint. Neq; solum, ubi unus numerus alteri, aut duo uni, uerum etiam, ubi duo duobus, aut tres uni æquales fuerint, nodum explicant. Hunc aût librum ideo seorsim edere placuit, ut hoc abstrusissimo, & plane inexhausto totius Arithmeticæ thesauro in lucem eruto, & quasi in theatro quodam omnibus ad spectandum exposito, Lectores incitarêtur, ut reliquos Operis Perfecti libros, qui per Tomos edentur, tanto auidius amplectantur, ac minore fastidio perdiscant.

The title page from Cardano's book Artis magnae sive de regulis algebraicis, which is usually referred to as Ars magna.

Cardano vs Tartaglia
Who was maligned?

"I am so pleased you accepted my invitation to come to Milan. I have so wanted to meet you and discuss your algebraic work. You have a rare talent," Cardano said, trying to flatter his guest.

"I am glad to be here. In your letters you have been most pressing to have us discuss my techniques for solving certain cubic equations," Tartaglia said coming directly to the reason for Cardano's invitation.

"Yes, my dear friend, but there is no rush. We have time," Cardano replied.

"The best time is the present," Tartaglia insisted having heard rumors about Cardano.

"As you wish. I have been entreating you to show me your methods. Revealing your findings would facilitate the

Girolamo Cardano (1501-1576)

progress of algebra," Cardano pointed out.

"I plan to publish them. I just have not had the time. Where do you find time to practice medicine and write so prolifically?" Tartaglia asked.

"I like diversity. There must be something I can do to convince you to share your findings. What do you want from me for your secrets of the cubic?" Cardano was now almost begging. "I'll give you my promise. No, I'll do more than that. I will take an oath on the Sacred Gospel to keep your discovery secret. Otherwise I will not be worthy of being a Christian or a gentleman. Can we agree on this as gentlemen?" Cardano asked.

"Since you put it in those terms, I would be willing to demonstrate my techniques," Tartaglia replied.

Cardano's persuasiveness won out once again. In March of 1539 Niccolò Tartaglia shared the work he had developed on solving certain cubic equations with Girolamo Cardano, but Tartaglia's explanation was evasive. Instead of explaining his method outright, he chose to present it in cryptic verses.

> **Quando che'l cubo con cose apresso**
> When the cube is brought closer to the things

> **Se agguaglia a qualche numero discreto**
> It becomes equal to some moderate quantity

> **Trovan dui altri, differenti in esso ...**
> Two others are found, different than it ...

Nevertheless, Cardano mastered and expanded Tartaglia's technique. Six years later Cardano broke his promise and published the results in a book.

The story of this scandal just begins here. Cardano led a very colorful and complicated life. There are numerous things that happened to him that helped form his character. Among these are:

—The illegitimacy of his birth kept him from practicing medicine in Milan for many years until his fame as a physician made the College of Physicians in Milan accept him in 1539.[1]

—He was a compulsive gambler, but even so he scientifically scrutinized games of chance. His *Liber de ludo aleae (A Book on Games of Chance)* 1663, the first book dealing with probability, was published posthumously.

—Omens and horoscopes influenced his decisions, and superstition played an important role in Cardano's life. He was an avid astrologer, who even worked out the astrological chart for Jesus Christ, which resulted in his being accused of heresy in 1570.[2]

—His elder son, who was an exceptional student and earned a degree in medicine, was involved in an unfortunate marriage. Apparently he could no longer put up with his wife, and poisoned her. This was a heavy blow to Cardano, since he was not able to save his son from the gallows in 1560.

Around 1542, Cardano hired a young man named Ludovico Ferrari as a servant. He quickly realized that Ferrari was very gifted,

Niccolò Tartaglia (circa 1500-1557)

and Cardano became his mentor and teacher. They worked on mathematics together, and during one of their sessions Cardano revealed Tartaglia's technique to the young man. Working together they expanded upon Tartaglia's initial work, and made a number of new discoveries. They knew, because of Cardano's promise to Tartaglia, they could not make their work public without revealing Tartaglia's work. Knowing that Tartaglia had first used his method to win a competition against a mathematician by the name of Scipio Ferro some thirty years earlier, they decided to research the archives. In 1543 they discovered in the writings of Ferro the same solution that Tartaglia had given Cardano. Since the technique appeared in Ferro's papers, Cardano no longer felt obligated to keep his oath. In 1545 Cardano published his algebra book *Ars Magna (Great Art)*. In the chapter of his book that dealt with cubic equations Cardano revealed Tartaglia's techniques along with the discoveries he and Ferrari had made. In the preface to this chapter, Cardano wrote that the chapter contained Tartaglia's method for solving certain cubic equations. He also provided the history behind the method giving credit to both Ferro and Tartaglia, and even the incident of his oath. In addition, he explained how he carried the work to subsequent steps.

Even though Cardano gave appropriate credit, Tartaglia was very upset when the book came out, accusing Cardano of being a thief, a scoundrel, and of breaking a sacred oath. Tartaglia continued his attack for many years. Ferrari took up the defense of Cardano by vehemently responding to Tartaglia's letters and challenging him to public debate.

Was Tartaglia justified in labeling Cardano a scoundrel and clouding his reputation for years to come?

---

[1]In one of his publications he brought to light the poor medical practices of Italian doctors, which was enthusiastically received by the public, but frowned upon by doctors. By the 1550s he had become a renowned doctor, having treated the Pope and the Archbishop of St. Andrew's in Scotland by diagnosing his allergy to feathers in his bed.

[2]Many famous people testified on his behalf, and the Church released him from prison. After the entire ugly affair was over, the Pope even granted him a pension.

Bibliography

Alic, Margaret. *Hypatia's Heritage*. Beacon Press, Boston: 1986.

Asimov, Isaac. *Asimov's Biographical Encyclopedia of Science & Technology*. Doubleday & Co. Inc., Garden City, NY: 1972.

Ball, W.W. Rouse. *A Short Account of the History of Mathematics*. Dover Publications, Inc., New York: 1960.

Barrow, John D.. *Pi in the Sky*. Clarendon Press, Oxford: 1992.

Bell, E.T.. *Men of Mathematics*. Simon & Schuster, New York: 1965.

Bernal, J.D.. *Science in History*. The MIT Press, Cambridge, MA: 1985.

Bernstein, Peter L.. *Against The Gods*. John Wiley & Sons, New York: 1996.

Boyer, Carl B.. *A History of Mathematics*. Princeton University Press, Princeton, NJ: 1985.

Clawson, Calvin. *The Mathematical Traveler*. Plenum Press, New York: 1994.

Dantzig, Tobias. *Number—The Language of Science*. The Macmillan Co., New York: 1930.

Davis, Philip J. & Reuben Hersh. *The Mathematical Experience*. Houghton Mifflin Company, Boston: 1981.

Dunham, William. *Journey Through Genius,* John Wiley & Sons, Inc., New York: 1990.

Dunham, William. *The Mathematica; Universe,* John Wiley & Sons, Inc., New York: 1994.

Eames, Charles and Ray. *A Computer Perspective,* Harvard University Press, Cambridge, MA: 1990.

Eves, Howard. *In Mathematical Circles,* Prindle, Weber & Schmidt, Inc., Boston: 1969.

Bibliography

Fauvel, John; Raymond Flood, Michael Shortland, and Robin Wilson, editors. *Let Newton Be!,* Oxford University Press, Oxford: 1989.

Goldstein, Thomas. *Dawn of Modern Science,* Houghton Mifflin Co., Boston: 1988.

Hall, Tord. *Carl Friedrich Gauss, a biography,* The MIT Press, Cambridge, MA: 1970.

Heath, Sir Thomas, translator. *Euclid's Elements,* Dover Publications, New York: 1956

Hodges, Andrew. *Alan Turing —the enigma,* Simon & Schuster, New York: 1983.

Hollingdale, Stuart. *Makers of Mathematics,* Penguin Books, London: 1989.

Hyman, Anthony. *Charles Babbage—Pioneer of the Computer,* Princeton University Press, Princeton, NJ: 1982.

Jones, Richard Foster. *Ancients and Moderns,* Dover Publications, Inc., New York: 1961.

Kline, Morris. *Mathematics — A Cultural Approach,* Addison-Wesley Publishing Co., Inc., Reading, MA; 1962.

Kline, Morris. *Mathematics and the Search for Knowledge,* Oxford University Press, New York: 1985.

Kline, Morris. *Mathematics in Western Culture,* Oxford University Press, Nw York: 1953.

Kline, Morris. *Mathematical Thought from Ancient to Modern Times,* Oxford University Press, New York: 1972.

Macrone, Michael. *Eureka! — What Archimedes Really Meant,* Harper Collins, New York: 1994.

McLeish, John. *Number—The history of numbers and how they shape our lives,* Fawcett Columbine, New York: 1991.

Bibliography

McLeish, John. *The Story of Numbers,* Fawcett Columbine, New York: 1994.

Newman, James R.. *The World of Mathematics,* Simon & Schuster, New York: 1956.

Osen, Lynn M.. *Women in Mathematics,* The MIT Press, Cambridge, MA: 1988.

Pais, Abraham. *Einstein lived here,* Oxford University Press, Oxford: 1994.

Palfreman, Jon and Doron Swade. *The Dream Machine,* BBC Books, London: 1991

Pappas, Theoni. *The Joy of Mathematics,* Wide World Publishing/Tetra, San Carlos, CA: 1989.

Pappas, Theoni. *More Joy of Mathematics,* Wide World Publishing/Tetra, San Carlos, CA: 1991.

Pappas, Theoni. *The Magic of Mathematics,* Wide World Publishing/Tetra, San Carlos, CA: 1994.

Perl, Teri. *Women & Numbers,* Wide World Publishing/Tetra, San Carlos, CA: 1995.

Regis, Ed. *Who Got Einstein's Office?,* Addison-Wesley Publishing House, Reading, MA; 1994.

Roan, Colin A.. *Science—Its History & Development Among the World's Cultures,* Facts on File Publications, New York: 1982.

Schwinger, Julian. *Einstein Legacy,* Scientific American Books, New York: 1986.

Singh, Jagjit. *Great Ideas of Modern Mathematics,* Dover Publications, Inc., New York: 1959.

Smith, D.E.. *History of Mathematics,* volume 1. Dover Publications, Inc., New York: 1951.

Bibliography

Smith, David Eugene. *A Source Book in Mathematics,* Dover
 Publications, Inc., New York: 1959.

Smith, David Eugene. Mathematics. Cooper Square Publishers,
 Inc., New York: 1963.

Struik, D.J., editor. *A Source Book in Mathematis, 1200-1800,*
 Harvard University Press, Cambridge, MA: 1969.

Struik, Dirk J.. *A Concise History of Mathematics,* Dover
 Publications, Inc., New York: 1967.

Swade, doron. *Charles Babbage and his Calculating Engines,* Science
 Museum, London: 1991.

Swetz, Frank J.. *From Five Fingers to Infinity,* Open Court,
 Chicago: 1994.

Wertheim, Margaret. *Pythagoras' Trousers,* Random House, New
 York: 1995.

Index

Mathematics teacher and consultant Theoni Pappas received her B.A. from the University of California at Berkeley in 1966 and her M.A. from Stanford University in 1967. She is committed to demystifying mathematics and to helping eliminate the elitism and fear that are often associated with it.

In addition to *Mathematical Scandals,* Pappas is the author of the following books: *The Joy of Mathematics, More Joy of Mathematics, Math Talk, Mathematics Appreciation, Greek Cooking for Everyone, Fractals, Googols and Other Mathematical Tales, The Magic of Mathematics, Music of Reason, Mathematical Footprints, Math-A-Day,* and *Math Stuff.* Her other innovative creations include *The Mathematics Calendar, The Children's Mathematics Calendar, The Mathematics Engagement Calendar, The Math-T-Shirt,* and *What Do You See?*—an optical illusion slide show with text.

Mathematics Titles by Theoni Pappas

MATH-A-DAY
$12.95 • 256 pages •illustrated•ISBN:1-884550-20-7

MATH STUFF
$12.95 • 232 pages •illustrated•ISBN:1-884550-26-6

MATHEMATICAL FOOTPRINTS
$10.95 • 156 pages •illustrated•ISBN:1-884550-21-5

MATHEMATICAL SCANDALS
$10.95 • 160 pages •illustrated•ISBN:1-884550-10-X

THE MAGIC OF MATHEMATICS
$12.95 • 336 pages •illustrated•ISBN:0-933174-99-3

FRACTALS, GOOGOL, and Other Mathematical Tales?
$10.95 • 64 pages • for all ages
illustrated•ISBN:0-933174-89-6

THE JOY OF MATHEMATICS
$10.95 • 256 pages
illustrated•ISBN:0-933174-65-9

MORE JOY OF MATHEMATICS
$10.95 • 306 pages
cross indexed with *The Joy of Mathematics*
illustrated•ISBN:0-933174-73-X

MUSIC OF REASON
Experience The Beauty Of Mathematics Through Quotations
$9.95 • 144 pages • illustrated•ISBN:1--884550-04-5

MATHEMATICS APPRECIATION
$10.95 • 156 pages • illustrated•ISBN:0-933174-28-4

MATH TALK
mathematical ideas in poems for two voices
$8.95 • 72 pages • illustrated•ISBN:0-933174-74-8

THE MATHEMATICS CALENDAR
$10.95 • 32 pages • written annually
illustrated • ISBN:1-884550-

THE CHILDREN'S MATHEMATICS CALENDAR
$10.95 • 32pages • written annually